LASERS AND OPTICAL
FIBERS IN MEDICINE

Physical Techniques in Biology and Medicine

Edited by

Denis L. Rousseau
AT&T Bell Laboratories
Murray Hill, New Jersey

William L. Nastuk
Columbia University
New York, New York

Denis L. Rousseau (ed.), *Optical Techniques in Biological Research*

Carleton H. Baker and William L. Nastuk (eds.), *Microcirculatory Technology*

Charles E. Swenberg (ed.), *Imaging Techniques in Biology and Medicine*

Abraham Katzir, *Lasers and Optical Fibers in Medicine*

LASERS AND OPTICAL FIBERS IN MEDICINE

Abraham Katzir

School of Physics and Astronomy
Tel Aviv University
Tel Aviv, Israel

Academic Press, Inc.

A Division of Harcourt Brace & Company

San Diego   New York   Boston   London   Sydney   Tokyo   Toronto

Academic Press, Inc.
1250 Sixth Avenue, San Diego, California 92101-4311

United Kingdom Edition published by
Academic Press Limited
24–28 Oval Road, London NW1 7DX

Library of Congress Cataloging-in-Publication Data

Katzir, Abraham.
    Lasers and optical fibers in medicine   /   Abraham Katzir.
        p.   cm.   --   (Physical techniques in biology and medicine)
    Includes bibliographical references and index.
    ISBN 0-12-401940-4
    1. Lasers in medicine.   2. Optical fibers in medicine.   I. Title.
II. Series.
    [DNLM: 1. Lasers--therapeutic use.   2. Laser Surgery.   3. Fiber
Optics.   4. Endoscopy--methods.   WB   117 K1951 1993]
R857.L37K38   1993
610' .28--dc20
DNLM/DLC
for Library of Congress   93-12744
                                                                                CIP

PRINTED IN THE UNITED STATES OF AMERICA
93 94 95 96 97 98   BB   9 8 7 6 5 4 3 2 1

Dedicated to the memory of my father,
Professor Aharon Katzir (Katchalsky),
A prominent scientist and a humanist,
A Pillar of Fire who gave Light to so many.

Contents

3   *Applications of Lasers in Therapy and Diagnosis*

4   *Single Optical Fibers*

5   *Optical Fiber Bundles*

6   Endoscopy

7   Fiberoptic Diagnosis

8 Fiberoptic Laser Systems for Diagnostics and Therapy

9   *Clinical Applications of Fiberoptic Laser Systems*

Appendix

Preface

Major developments in modern optics and electrooptics occurred in the late 1950s and the early 1960s. These have had great impact on medicine. The development of optical fibers led to the development of the endoscope and to endoscopic imaging and therapy. With the discovery of the first laser by T. Maiman in 1960 came the realization of its potential as a useful tool in the hands of physicians. Indeed the Ar, Nd:YAG, and CO_2 lasers have been widely used in medicine ever since. During that early period there were three groups of users of these tools: the scientists who used the new "toys" for scientific research, the industrial or military users who did their own research and development, and the clinicians who bought lasers and used them for laser surgery or therapy. There was practically no interaction among the three groups. Each group had its own meetings, journals, "slang," and research. Even now, one is unlikely to find medical papers at physics meetings or modern optics studies at medical meetings.

Things started to change in the mid-1980s, when it became apparent that further progress depended on a close collaboration among researchers in various interdisciplinary fields. Gradually researchers started working in larger teams and participating in joint conferences and symposia. Over the years the whole area of biomedical optics dealing with lasers, fibers, and modern optics in medicine has emerged as a distinct discipline.

Over the past 10 years I have been involved in organizing interdisciplinary symposia in this field. I am indebted to SPIE, to Joe Yaver, its executive director, and to the SPIE directors, presidents, and staff for giving me this opportunity. The idea of writing this book stemmed from my interaction with the researchers who participated in the Biomedical Optics Symposia.

The main goal of the book is to provide a basis for the understanding and use of lasers and optical fibers in medicine. The principles of the operation of lasers are discussed first, with emphasis on the lasers that are commonly used in medicine. This is followed by the principles of the different types of interactions of various laser beams with human tissue, with emphasis on the special uses of lasers in diagnosis, therapy, and surgery. The principles of operation of single optical fibers are then presented. This is followed by a description of the operation of bundles of fibers that are used for illumination and for endoscopic imaging inside the body. Lasers and optical fibers may be integrated into systems that provide imaging, diagnosis, and therapy inside the body; some of these systems are described. Finally, the applications of laser and optical fiber systems in specific medical disciplines, including cardiology, gastroenterology, general and thoracic medicine, gynecology, neurology, oncology, ophthalmology, orthopedics, otolaryngology, and urology, are discussed in detail.

A special effort has been made to enable researchers from various fields to understand and use this book. Toward this goal each chapter and topic is presented in a three-tier manner, as described in the prologue. The book should be useful as a source for scientists, engineers, and physicians, and as a text in courses in medical schools and departments of biomedical engineering.

I have received much help from many physicians, scientists, and engineers in the course of my work. I am grateful to the following friends and colleagues, who generously gave their time to read the whole manuscript and provide enlightening remarks: Tom Deutch, Ari DeRow, Frank Frank, Jim Harrington, Steve Joffe, Betty Martin, Halina Podbielska, Ofer Shoenfeld, Kevin Shoemaker, Larry Slifkin, Johannes Tschepe, and Joseph (Jay) Walsh. Special thanks go to Frank Cross and Frank Moser, who read the different versions of the complete text and made invaluable contributions.

In the course of preparing the book I have corresponded with almost 100 individuals and companies who sent me photographs and illustrations. I thank all of them and, in particular, those whose material is included: Advanced Interventional Systems, Candela, Coherent Inc., Dr. Elma Gussenhoven, Dr. Basil Hirschowitz, Dr. Steve Lam, Laser Diode Laboratories, Laser Industries, Laser Sonics, Laser Surgical Technologies (LST), Luxtron, The Medical Library of the University of Vienna, Dr. Ted Maiman, Mitsubishi Cable Industries, Olympus Corporation, PDT Systems, Pentax Corporation, Schwartz Electro Optics, Storz, Dr. Yasumi Uchida, and Dr. Rudy Verdaasdonk.

I thank my associates at Tel Aviv University: Herman Leibowitch and Yoram Weinberg for their photographic assistance; Benni Bar, Abraham Yekuel, and Arie Levite for their devoted help; and Ninette Corcos for her expert drawings used in this book.

I thank my editors at Academic Press, Charles Arthur, Steven Martin, and Marvin Yelles, for their contributions and for their continuous encouragement and support.

Last, but not least, I express my gratitude to my family: my wife and close friend, Yael, who shared with me this long voyage; my children, Dan and Tammy, who bore with me and supported me along the way; and my mother, Rina, whose strong spirit gave courage to all of us.

Abraham Katzir

Prologue

The field of fiberoptics in medicine is an interdisciplinary one, involving science, engineering, and medicine. Some of the researchers in this field are scientists interested in the physics of optical fibers; others are chemists interested in the optical triggering of certain chemicals. Engineers may be attracted to the challenge of the new problems in designing and making laser–fiber systems. Some physicians are interested in the more scientific aspects of laser–tissue interactions and in animal experiments; others are more interested in the implementation of these techniques on patients. In organizing this book the various interests of these groups were taken into consideration, and therefore most chapters are divided into three basic sections. These are arranged in a three-tier system:

(i) Fundamentals: This section describes the basic concepts of the topic. It gives the reader who is not well versed in lasers or fiberoptics the necessary background for understanding the various phenomena, without the complex scientific details.

(ii) Principles: This section is for the reader who is interested in a more "in-depth" treatment of a given topic. The details are still kept to the more practical aspects of a problem; however, a more comprehensive engineering approach is given.

(iii) Advances: More scientific details are given in this section. Although not necessary for understanding the problems in general, the scientific details are often vital for researchers and for those interested in the finer details.

People who have not yet been exposed to the fascinating fields of lasers and fiberoptics should read the Fundamentals sections during the first reading of the book. It is only through subsequent use of the book that the more detailed sections will become pertinent. Those with sufficient background are encouraged to read the Principles and the Advances sections as well.

1

Introduction

1.1 HISTORICAL BACKGROUND

Light has been used by physicians for both diagnostic and therapeutic procedures since the dawn of civilization. Observation being the only diagnostic tool, light enabled them to see skin color, inspect eyes or wounds, and then choose a suitable course of therapy. Heat from sunlight, or even from light emitted from campfires, was used for therapy throughout the ages. Both the Greeks and the Romans used to take daily sunbaths and a solarium was part of many Roman houses. Illumination by sunlight was used in ancient Egypt for therapeutic application in skin diseases.

Even during the early days of medicine it was clearly understood by physicians that they would benefit enormously if they could diagnose and treat the inside of the body using nonsurgical medical instruments. One of the first attempts was the development of a simple optical instrument to look inside the body. A tube inserted into the ear or the mouth enabled some limited view inside the body and was called an endoscope. The word endoscope stems from the Greek *endo*—within, and *skopien*—to view. Over the last few hundred years these instruments have been much improved mechanically and optically. With the availability of optical fibers and lasers, the endoscope became a more complicated but much more powerful diagnostic as well as therapeutic tool. The new fiberoptic endoscopes and "integrated" systems such as the laser catheter or the laser endoscope were to cause a revolution in many fields of medicine. If one can position optical fibers at a desired spot in the body, these systems cover virtually all the medical procedures, such as diagnosis, surgery, and therapy, with greatly reduced trauma to the patient. This book deals with this rapid expansion of laser and fiberoptic technology in medicine.

By way of introduction, the various stages that led to the development of the modern endoscopic systems are discussed (Berci, 1976; Haubrich, 1987). The *first phase* of endoscopy is based on hollow tubes and rudimentary optical elements. Many centuries before Christ, simple open tubes were used for the examination of body cavities. It was only in the early 1800s that Bozzini added improved illumination by using a wax candle light source and a 45° mirror to reflect light into the tube. Desormeaux made further improvements in 1867 by replacing the candle with an alcohol flame and the mirror with a lens. This type of rigid endoscope was inserted for the first time in 1868 by Kussmaul through the mouth, pharynx and esophagus, making it possible for the inside of the stomach to be seen. It is said that a professional sword swallower volunteered to be the first patient. This was a first attempt at gastroscopy. A few of the very early endoscopes are shown in Fig. 1.1.

Later in the 1800s there were further developments, such as the construction of several tubes which were telescoped into each other to facilitate easier introduction into the body. An element of flexibility was added by building part of the instrument with metal rings rather than from a rigid tube. In this first stage of endoscopy, the organ inside the body was illuminated through the tube and the physician looked directly at the organ through the tube. Another stage in the development of endoscopes was the use of a simple optical system for transmission of a full image from inside the body to the physician's eyes. One of the pioneers of this was Nitze, who in 1879, in collaboration with the instrument maker Leiter and others, placed a lens into an open tube and used a glowing platinum wire as a light source. He later used a miniature electric globe for this purpose (Lewan-

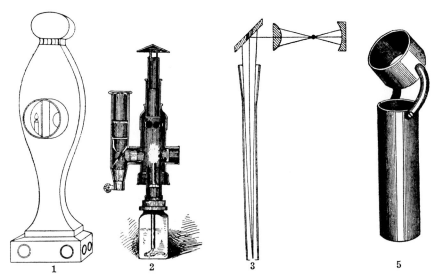

FIGURE 1.1   Earliest endoscopes: (1) Bozzini's apparatus; (2) Desormeaux's illuminating apparatus; (3 and 5) Kussmaul's and Bevan's esophagoscopes. (Courtesy of the Medical Library of the University of Vienna.)

dowski, 1892). At the same time Edison was making his improved electric bulbs, which were used by Mikulicz in 1881 to build better gastroscopes. Similar instruments were being made for use in other areas of medicine, such as the bronchoscope or cystoscope, and were gradually improved throughout the early 1900s. One of the improvements included a rubber tube, which was introduced first and served as a guide for the rigid endoscope. Another was an endoscope whose distal end was movable. It should be noted that as early as 1890, these endoscopists were using still cameras to record their findings. Musehold, for example, had photographed the larynx in 1893. A few of the endoscopes of this period are shown in Figs. 1.2–1.4.

The *second phase* in the development of endoscopes was the use of a series of lenses, rather than a single element, to transmit images. An important development was made by Schindler and Wolf in Germany in 1936, when they built a

FIGURE 1.2 Leiter rigid endoscopes with bent tip. (From Lewandowski, 1892. Courtesy of the Medical Library of the University of Vienna.)

FIGURE 1.3 A porter carrying the Nitze–Leiter cytoscope (1879). (Courtesy of the Medical Library of the University of Vienna.)

semiflexible gastroscope, consisting of a series of lenses. A further advancement was implemented by H. H. Hopkins in the 1940s in Britain by using a novel rod–lens system which produced a better-quality image. Hopkins-type endoscopes are still widely used today, as shown in Fig. 1.5.[1] During the first half of the 20th century, the rapid development of photographic technology was incorporated into endoscopy. First, color photographs were taken during gastroscopy, and later motion pictures were taken of the larynx, the esophagus, and the bronchial tree.

The *third phase* in the development of endoscopy was the introduction of fiberoptic endoscopes. These are based on thin and transparent threads of glass, the so-called optical fibers. The idea of transmitting images by using bundles of optical fibers dates back to the years 1927–1930 and to the independent work of J. L. Baird in the United Kingdom, C. W.Hansell in the United States, and H. Lamm in Germany. The idea lay dormant for two decades until it was revived independently in 1954 by A. C. S. van Heel in Holland and by H. H. Hopkins and N. S. Kapany in England. These researchers demonstrated that a bundle of thin optical fibers could be used to transmit a good-quality image, and with the improved flexibility of the fiber bundle the potential applications for optical fibers in

---

[1] A cross section of such an endoscope is also shown in Fig. 8.8.

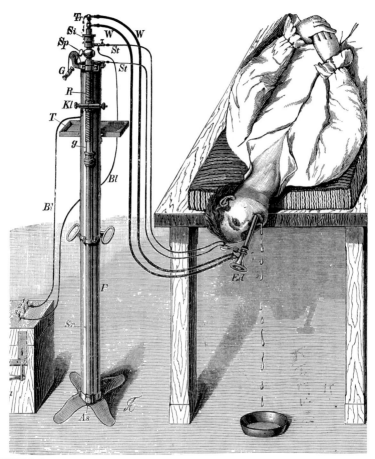

FIGURE 1.4 Mikulicz gastroscope. (From Lewandowski, 1892. Courtesy of the Medical Library of the University of Vienna.)

medicine could be greatly expanded. In 1959 the first medical instrument, a flexible fiberoptic gastroscope (Hirschowitz, 1979), was developed and used on patients by a physician (Hirshowitz) working in conjunction with two physicists (Curtiss and Peters). An early endoscope and an image transmitted through it are shown in Fig. 1.6 and Fig. 1.7.

Soon afterward, other fiberoptic endoscopes followed, such as bronchoscopes and colonoscopes. In these flexible endoscopes, the fiberoptic system provided both the illumination and the image transmission. Most of the endoscopes incorporated open (ancillary) channels through which the physician could inject saline solution or drugs or insert instruments such as forceps or electrosurgical instruments. An early commercial endoscope and a picture taken through it are shown in Fig. 1.8 and Fig. 1.9.

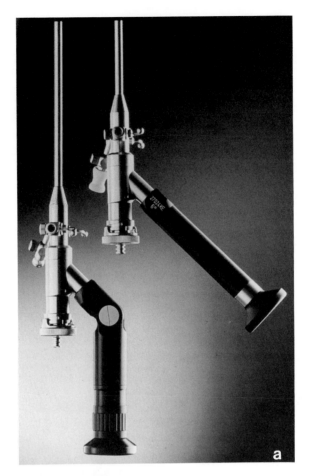

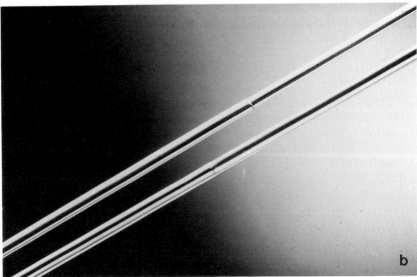

FIGURE 1.5 Modern (Hopkins-type) rigid endoscopes: (a) proximal and (b) distal ends. (Courtesy of Storz.)

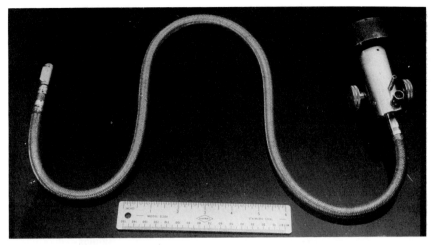

FIGURE 1.6   An early fiberoptic endoscope. (Courtesy of B. I. Hirschowitz.)

FIGURE 1.7   Photograph of a stamp taken through the endoscope pictured in Fig. 1.6. (Courtesy of B. I. Hirschowitz.)

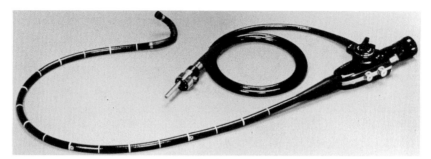

FIGURE 1.8   Early commercial fiberoptic gastroscope. (Courtesy of Olympus.)

In the early 1970s, there was rapid development in the field of optical fibers for communications. Optical fibers of superb quality and excellent transmission were produced for application in communication networks. Applied to endoscopy, these fibers brought about huge improvements; for example, thin endoscopes were developed for cardiology (angioscopes). These endoscopes have an outer diameter of less than 3 mm and incorporate more than 10,000 individual fibers in their imaging bundle to greatly enhance the optical resolution. They also include an ancillary channel for irrigation or for insertion of an additional instrument. These

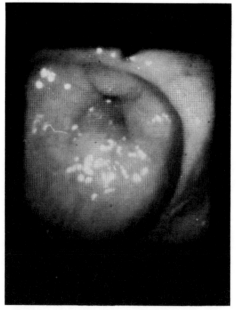

FIGURE 1.9   Picture of a stomach taken through the endoscope pictured in Fig. 1.8. (Courtesy of Olympus.)

thin instruments enable physicians to obtain images from areas which were, until now, quite beyond their reach.

In 1960, the first laser (the ruby laser) was developed by T. Maiman. This new source of light has many technological and military applications. Ever since the first laser was developed, it was believed that lasers would have a far-reaching impact on medicine. In particular, three properties were recognized as extremely important: (a) the laser emits a collimated (parallel) beam of light, which can be very powerful; (b) the light is practically monochromatic (consisting of one color); and (c) one of the unique features of laser light, known as coherence, makes it possible to focus it onto a very small spot. The energy density (energy divided by irradiated area) in this focal spot is high, and when the light is absorbed in biological tissue it can cause a rapid rise in temperature. At low energy levels, the heat generated in the spot coagulates blood and causes hemostasis (bleeding ceases) of small blood vessels. At higher levels, it causes vaporization and removal of tissue. These are the basic phenomena used in laser surgery. During the four years following the invention of the first laser, three other lasers were invented: the Ar (argon) ion, the CO_2 (carbon dioxide), and the Nd:YAG (neodymium : yttrium aluminum garnet) lasers. These are still the most widely used lasers in medicine and surgery. In 1963 the CO_2 laser was applied in dermatology for tissue removal. This was followed in 1964 by a more spectacular success: the use of the ruby laser beam for "noninvasive" operations inside the eye. This red laser beam was transmitted through the transparent compartment of the eye and focused on the highly vascularized retina for photocoagulation.

During the last 25 years, there has been continuous growth in the use of lasers for a large variety of medical applications. They are based on various types of lasers and different types of laser–tissue interactions. Some are briefly mentioned below:

(I) Diagnosis

   Low-intensity lasers interacting with tissue may give rise to a characteristic luminescence which supplies information regarding the tissue, such as blood flow, pH, and oxygen content. They can also indicate pathological alterations in the tissue caused by cancer.

(II) Therapy

   (i) Photochemotherapy: The laser light triggers chemical reactions in the body which are utilized for therapeutic applications.

   (ii) Heating—laser coagulation and welding: Blood is coagulated by the heat generated by the laser energy, thus sealing blood vessels of small diameter. This is useful for controlling bleeding during laser surgery. The heat generated by the beam can also be used to "weld" tissue or anastomose together two segments of tissue instead of using sutures. This is applied for attaching tubular organs such as intestines or blood vessels and for closure of incisions.

(iii) Vaporization—laser surgery: Lasers can be used to cut through tissue, by vaporization, in such areas as general surgery, gynecology, orthopedic surgery, and neurosurgery.

(iv) Pulsed laser effects: With the availability of lasers which emit energy in very short bursts (nanoseconds to milliseconds range) of enormous peak power (megawatts), new effects are possible. These short pulses are used to remove tumors inside the eye, fragment urinary or gallbladder stones, or cut very hard tissue such as bone or dentin.

Lasers have been used in the treatment of more than 1 million patients worldwide. Although the use of lasers in medicine is certainly growing, in most cases lasers are being employed simply to supplement conventional techniques, such as using a laser as a scalpel. In some of these cases, there may be no real advantage to using the laser and conventional techniques are either less expensive or faster. With the development of optical fibers, the situation has suddenly changed. The fiber can be inserted inside the body and the laser beam that is transmitted through it interacts with tissue. The fiber can be inserted into the body via a flexible plastic tube—a regular catheter. This combination of laser + fiber + catheter is called a laser catheter. Alternatively, an extra fiber can be inserted into the ancillary channel of an endoscope. This combination of laser + fiber + endoscope is called a laser endoscope and constitutes the *fourth phase* of endoscopy. The physician can insert the whole endoscope inside the body, obtain an image through the endoscope, and, if desired, transmit a laser beam through the extra fiber for therapy or for diagnosis. The laser catheter or the laser endoscope is inserted into the body via the natural openings, through a small incision in the skin or even via a large hypodermic needle. Thus, even if the laser can do no better than conventional surgery, its mode of delivery is less invasive and it reduces the need for major operations.

1.2 ILLUSTRATIVE EXAMPLES

The potential use of these new techniques is illustrated by a few examples which are discussed in the ensuing chapters.

The first example deals with the major problem of heart disease. One of the leading causes of death in the Western world is acute myocardial infarction. The mechanism which leads to this condition is the buildup of atherosclerotic plaque which gradually occludes the coronary arteries that supply the heart. The patency (openness) of the arteries is analyzed by an x-ray method called angiography. This method relies on injecting an x-ray–opaque liquid into the arteries through a catheter and studying the x-ray shadow of this liquid. If the artery is obstructed, the cardiologist tries to "remodel" the artery (if the patient has clinical symptoms). The cardiologist inserts a catheter which contains a balloon at its distal end and threads it through the blockage. The balloon is inflated and the opening enlarged by compressing the plaque to the sides of the artery. This method is sometimes called balloon angioplasty, but its more accurate name is percutaneous transluminal coronary angioplasty (PTCA). It can be used only if there is an adequate opening in the blockage that allows passage of the balloon. In about a third of cases, the remodeled or recanalized artery closes again (restenosis) within a few months. In cases of severe blockage, open heart surgery has to be performed—a "bypass" operation. In this procedure, another blood vessel is placed parallel to the diseased artery (thus bypassing the blockage) and blood flow is improved. This is a major operation that is traumatic, carries risk, and is costly for both the patient and society. There are many cases in which PTCA is either not applicable or inadequate to solve the problem. In some of these cases the laser catheter has already been used instead. The fiber is inserted into the circulation and threaded into the coronary artery. X-ray methods help in guiding the catheter to its proper position. A laser beam transmitted through this fiber is then used to ablate the atherosclerotic plaque and restore blood flow in the artery. It is expected that the plaque might be selectively vaporized, without causing thermal damage to the blood vessel wall and without leaving debris. It is also hoped that this procedure will not result in unacceptable short-term restenosis. The use of the laser catheter is obviously not limited to the coronary arteries; it may be used in peripheral arteries as well. It could be applied to vaporize blood clots or to perform microsurgery. Extensive clinical studies are already under way using this technique.

A second example of laser application is the treatment of urinary stones. Cystolithotomy, the surgical removal of bladder stones, is one of the oldest surgical operations (others are circumcision and trephination of the skull) and has been performed for thousands of years. Hippocrates, in the fifth century B.C., states in his oath, "I will not use the knife, not even on sufferers from stone, but will withdraw in favor of such men as are engaged in this work." Various surgical techniques of stone removal have been mentioned in the medical literature, as far back as Celsus in the first century and continuing through the present. In France in 1824, Civiale developed an instrument which could be inserted through the urethra and used to drill holes in stones located in the bladder. This procedure replaced the "cutting of the stone." One of the major problems of this method was that it had to be done blindly. This was solved by Nitze in 1877 by introducing a rigid cystoscope which was used for imaging inside the bladder. This endoscope paved the way for modern developments of urinary tract therapy and surgery. With the development of flexible instrumentation in the 1980s, a laser catheter was inserted into both the urethra and the ureter. A laser beam transmitted through the fiber was successfully used to disintegrate urinary stones. Since then, this laser lithotripsy procedure has been widely used.

One of the disadvantages of the laser catheter is the inability of the physician to get a clear view of the area which is being treated. This can be overcome by using the laser endoscope. This instrument is used in cancer treatments such as photodynamic therapy (PDT), which is based on the activation of drugs by use of laser light. A chemical, such as hematoporphyrin derivative (HPD), is injected into the body. This drug has three characteristics: (i) after several hours, the HPD accumulates selectively in malignant cells; (ii) if HPD is then illuminated with ultraviolet (UV) or violet light, it emits a characteristic red light; and (iii) if HPD is illuminated with intense red light, it causes the release of an agent (singlet oxygen) which kills the malignant host cells. Suppose that a physician performs a regular endoscopic examination inside the body and discovers a suspicious tumor. HPD is injected into the body and after a couple of days the physician inspects the tumor, while illuminating it with the UV light rather than the regular white light. The physician now looks at the tumor through the endoscope using a red filter, and if the tumor emits a red light, it is malignant. If the physician now illuminates the same tumor with a high-intensity (laser) red light through the same endoscope, the HPD is activated. Singlet oxygen forms and the tumor cells are killed. This method has been tried clinically on thousands of patients, but it is still experimental. Yet it illustrates how a laser endoscope may be used for both diagnosis and therapy.

Three of the main causes of death in the Western world are heart diseases, cancer, and cerebral diseases. In most of these disease the lesions (the altered organ or tissue whose function is impaired) are at sites that cannot be observed directly. Laser endoscopy is already used for the diagnosis and therapy of these diseases. Other areas in which this technique is used include gynecology, otolaryngology, orthopedic surgery, gastroenterology, general surgery, neurosurgery, and urology. In many of these cases, laser endoscopy will replace the traditional procedures that often involve major surgery. Clinical studies have been carried out in all these areas. It should be stressed again that this modality is the least invasive method of treatment and may often be adequately performed in the physician's office. The combined laser and fiberoptic techniques will cause a fundamental change in all these medical disciplines.

1.3 THE BOOK STRUCTURE

The general structure of the book is as follows:

In Chapter 2, lasers and in particular "medical" lasers are discussed. A brief description of a laser and the unique characteristics of laser light is given. Included in this chapter is a description of the lasers that are widely used in medicine today or that seem to be potentially important. Gas lasers such as CO_2, Ar, and excimer lasers; solid-state lasers such as Nd:YAG or Er:YAG; tunable lasers such as the liquid dye lasers; semiconductor lasers such as GaAs; and novel additions such as the free-electron laser are discussed. Characteristics such as wavelength or power output are described for each laser. Also included in this chapter are some of the principles of laser beam optics, such as the characteristics of the beam itself, the guiding of the beam by mirrors, and focusing of the beam by lenses. At the end of this chapter the problem of laser safety is briefly discussed.

Chapter 3 describes the uses of lasers for therapeutic and diagnostic purposes. Some of the scientific principles of the interaction between laser beams and materials, such as transmission of laser beams through materials or absorption of laser energy in materials, are discussed in general terms. This chapter also describes the laser–tissue interaction, which varies from tissue to tissue, depending on both the tissue involved and the wavelength, temporal behavior (pulsed or continuous-wave laser), and spatial characteristics (power distribution) of the laser beam. Different types of interaction suggest different clinical applications. The use of lasers for diagnostic purposes and the scientific principles and clinical applications are discussed. Laser therapy can be divided into two areas, one based on thermal and one on nonthermal effects. Included under thermal effects are laser coagulation, laser welding of tissue, and laser surgery. Nonthermal effects are the triggering of biochemical reactions and short-pulse laser tissue photoablation. All these phenomena are important for understanding laser diagnostics and laser therapy using optical fibers and fiberoptic systems.

Chapter 4 discusses the optical fibers that are used in medicine, including the fundamentals of light transmission through single optical fibers. The more commonly used fibers made of silica-based glasses are described, as well as special fibers for the transmission of UV, visible, and infrared (IR) radiation and the transmission of high-power laser beams.

Chapter 5 deals with arrays of fibers and fiberoptic bundles that provide the basis for illumination and imaging systems. A description is given of the operation of two types of bundles of fibers: nonordered bundles for illumination and ordered bundles for image transmission. An explanation of the fabrication and the properties of fiberoptic imaging systems or fiberscopes follows. This chapter lays the scientific foundation for understanding the clinical applications of the systems which are described in later chapters.

Chapter 6 describes fiberoptic endoscopy and its applications. Medical instruments based on imaging bundles of fibers, endoscopes, are discussed in some detail. The chapter deals mostly with nonlaser applications of endoscopes, including imaging inside the body. There are numerous other applications which involve the insertion of instruments through the ancillary channels. Mechanical tools such as forceps or cutting tools are sometimes used, as well as electrical instruments and microwave antennas. A few special techniques such as fluorescence imaging, video endoscopy, and ultrasound imaging are also described. The commonly used endoscopes and their advantages are discussed in general and are illustrated with pictures and data from clinical uses of endoscopy in medicine.

Chapter 7 describes the use of optical fiber systems for diagnostic purposes. Light transmitted through an optical fiber impinges on blood or tissue. Emitted luminescence or reflected light is transmitted back through an optical fiber into an analyzing system. The optical analysis of the signal sent back from the body can serve to perform a variety of diagnostic measurements. Using such techniques, physicians are able to monitor blood pressure and temperature inside the heart as well as gas content, pH, or glucose concentration in the blood. These measurements are carried out endoscopically in real time. Instead of extracting blood from patients and sending it for analysis in a laboratory, the analysis may be performed in the operating room or in the physician's office. This procedure increases the possibilities of using fiberoptic systems for a large variety of diagnostic purposes.

Chapter 8 addresses fiberoptic systems for therapy. It describes several of the integrated medical systems that comprise lasers, endoscopes (or catheters), and "power" fibers for the delivery of laser power. One part of this chapter deals with laser catheters and their applications and describes some of the systems that are presently under investigation. Some of the studies that are currently being done *in vivo* and *in vitro* are described. In the second part of this chapter, laser endoscope systems are described. Again, the systems that are currently being studied *in vitro* and *in vivo* are discussed, as well as some of the laser endoscopic systems that are widely used.

Chapter 9 details some of the clinical applications of laser–fiberoptic systems in several medical disciplines, such as cardiology, gastroenterology, general surgery, gynecology, neurosurgery, oncology, ophthalmology, orthopedics, otolaryngology, and urology.

An Appendix which contains some supplementary material, a detailed Bibliography of books that are relevant to our topic, and an extensive Glossary of terms and abbreviations are included at the end of the book.

References

Berci, G. (1976). History of endoscopy. In: Berci, G. (Ed.), *Endoscopy*, pp. xix–xxiii. New York: Appleton-Century-Crofts.

Haubrich, W. S. (1987). History of endoscopy. In: Sivak, M. V. (Ed.), *Gastroenterologic Endoscopy*, pp. 2–19. Philadelphia: Saunders.

Hirschowitz, B. I. (1979). A personal history of the fiberscope. *Gastroenterology* **764**, 864–869.

Lewandowski, R. (1892). *Elektrische Licht in der Heilkunde*. Wien: Urben & Schwarzenberg.

2

Medical Lasers

2.1 INTRODUCTION

The last few decades have witnessed increased use of optical methods for medical diagnosis and treatment. For example, blood is analyzed by spectroscopic methods that involve absorption, reflection, or emission of light. In therapy, special heating lamps are sometimes used to relieve pain. Ordinary light sources such as incandescent lamps, fluorescent lamps, and other specialized light sources are used in all these applications.

In 1960, a totally new light source appeared—the laser. The word laser had already appeared in the writing of Plinius, the famous, first century A.D. historian. "The laser is numbered among the most miraculous gifts of nature and lends itself to a variety of applications" (Plinius, XII, 49). Plinius' laser was an herbal plant that grew on the shores of the Mediterranean and was used by the Romans for therapy. The modern word laser is an acronym for light amplification by stimulated emission of radiation. Such terms are recognized as part of the language of the field, called quantum electronics. The laser itself, however, can be explained in simple terms so that its operation and unique characteristics can be understood by users.

In the early 1960s, it was presumed that the laser possessed enormous potential for the field of medicine. One of the first medical applications was the treatment for a detached retina. A laser beam was directed into the eye, producing burn scars that were used to "weld" the retina to the inner wall of the eye. Laser beams of higher intensity were used to coagulate blood, and laser beams of even higher intensity were used for cutting tissue, that is, for laser surgery. The early laser systems were of very poor quality. Designed and utilized by engineers and scientists for uses other than medicine, they did not address the particular needs of medical applications. The lack of communication between physicians and engineers also hindered the development process. Only after two decades were good and reliable laser systems for medical purposes produced.

With the development of better-engineered lasers, the applications of lasers in medicine increased. The progress was slow but constant. Not only the number of laser operations but also the number of medical disciplines and specialists that utilized lasers increased. The laser industry made a great effort to educate physicians about lasers and their applications, which helped to introduce and incorporate the laser into medicine.

The transition of any new technology into a widely used practical system takes many years. This is true even for simple systems such as toasters, ovens, or dishwashers. If an analogy is drawn between lasers and cars, then the lasers of today are probably closer to the Model T Ford than to a modern car. Given time and more widespread use, they are bound to improve. During the early days of the car's development, the driver had to know quite a bit about the car, including its structure and limitations, in order to drive (and sometimes to survive). Even in modern cars, with computerized braking systems and navigational accessories, the driver must still learn how to operate the car. In many countries, the driver is even tested on the principles of the engine's structure. The driver must also remember to fill the radiator with water and check the pressure in the tires. The driver who understands how the braking system and the various controls work generally becomes a better driver.

Operating the laser is more difficult than operating and understanding a car, but even more important, the struggle to understand and control the interaction between lasers and human tissue is still taking place. In the author's opinion, it is essential that medical personnel using lasers be educated in the principles of laser science. It is even more important for the research physician to understand these topics. Without this knowledge, neither the clinician nor the research physician can put a laser to good and safe use.

2.2 LASER PHYSICS

2.2.1 Laser Physics—Fundamentals

2.2.1.1 Ordinary Light Sources

The term "light" refers to the visible part of the electromagnetic spectrum. It is also loosely used to refer to the infrared (IR) and the ultraviolet (UV) parts of the spectrum. There are many sources of light, such as the sun, incandescent lamps, and fluorescent lamps. For our purposes, we call these ordinary light sources and define and specify them in terms of the following parameters:

(i) Spectral composition (color): Ordinary light sources emit light of many colors and are called polychromatic (many colors). Light can be represented as waves and the different colors correspond to waves of different wavelengths (see Section 10.1). Some ordinary light sources have a limited range of wavelengths and appear to be red or blue or yellow (i.e., some definite color). Other light sources emitting many wavelengths appear to be "white." In each case, a range of wavelengths is involved.

(ii) Direction: The light of an ordinary lamp is emitted in all directions. It usually comes from a fairly large area, such as a glowing filament, or from discharge inside a fluorescent lamp.

(iii) Power: Power is the ratio between energy (usually measured in joules) and time (measured in seconds). Power is measured in watts, defined as joules/ second. For an ordinary light source, such as an incandescent lamp, the electric input power (typically 60 or 100 W) is given. The light source gives out light and heat as output power. The efficiency of most lamps is very low, in the sense that only a fraction of the electric power consumed by the lamp is converted into light. For a powerful lamp consuming 500 W (i.e., input power), the light emitted (i.e., output power) is only a few watts.

(iv) Order: When waves are emitted by a regular light source, in different directions, there is no correlation between the various waves (i.e., no order). The light source is thus termed incoherent.

(v) Power density: The output light power is usually distributed over the whole sphere around the lamp. The ratio between the emitted light power and the illuminated area is called power density (irradiance). For example, for the 500-W lamp discussed above, the power density may well be only 1 milliwatt per square centimeter at a distance of 10 cm from the lamp.

It should also be mentioned that only a small fraction of this power can be gathered, using a focusing lens or a mirror, and that the focal spot does not have a small area. Therefore even the illumination of this focal spot area is characterized by a low power density.

Figure 2.1 shows a comparison between a laser and an ordinary light source.

2.2.1.2 Laser Light

The laser is a completely different type of light source (Winburn, 1987). Three of its intrinsic characteristics are:

(i) Monochromaticity: One "color" (one wavelength) or more accurately a very narrow band of wavelengths is emitted by the laser. Such pure colors are not normally observed in nature. Although lamplight or sunlight that is passed through a colored glass filter produces "one color," this color corresponds to a relatively wide band of wavelengths. It is much less "pure" in this sense than laser light. The filter blocks out much of the energy and thus the power transmitted by the filter is low. By contrast, a laser inherently emits only monochromatic light, with all the power concentrated at a unique wavelength.

(ii) Collimation: Laser light is emitted in a beam that is quite narrow and stays narrow. This property makes it possible to send a laser beam from the earth

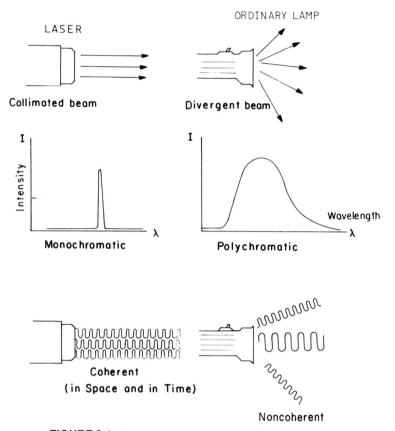

FIGURE 2.1   Comparison of a laser and an ordinary light source.

to the moon with very little divergence. For many practical purposes, this is a parallel beam.

(iii) Power: Power emitted by lasers may be quite high. For example, the *output* (light) power normally specified in medical laser beams can easily be 100 W or more.

Two important properties of the laser light that are not independent of (i)–(iii) should also be mentioned:

(iv) Coherence: As stated earlier, the collimated laser beam consists of waves of one color (i.e., waves of equal length). Coherence is a term used by optical physicists to express the degree of monochromaticity and collimation. All the waves in the laser beam are highly ordered in space and correlated in time. An analogy to explain this state of order is to relate the laser source to a large group of soldiers walking on a parade ground, with the same pace at the same time. An

TABLE 2.1   Laser versus Ordinary Light Source

Property	Laser	Ordinary light source
Directionality	Collimated (parallel beam)	Noncollimated (light emitted in all directions)
Color	Monochromatic (one color) Comment: coherent beam (i.e., ordered in time and space)	Polychromatic (many colors) Comment: noncoherent beam (i.e., nonordered)
Power output	Can be high	Medium or low
Temporal	Can produce very short and energetic pulses	Typically long and low-energy pulses
Power density	High; can be focused to a very small spot (of diameter $d = \lambda$)	Low; relatively large focal spot

ordinary light source is similar to the parade ground after the dismiss order has been given.

(v) High power density: Laser radiation is concentrated in a narrow pencil of light whose area is a few square millimeters. The power density of the laser beam itself is high. Moreover the beam can be focused onto a spot whose diameter is of the order of the laser wavelength. The power density at the focal spot is extremely high—many orders of magnitude higher than in the case of an ordinary lamp.

Table 2.1 shows the properties of laser light in comparison to ordinary light. All these properties of the laser are pertinent to the unique uses of the laser in medicine. Before explaining these points, the operation of a laser is briefly described.

2.2.1.3 Operation of a Laser

The fundamentals of a laser and how it works can be understood by an analogy between lasers and other systems which are better known to us. A basic system, found in most homes, is the regular stereophonic system, as shown in Fig. 2.2. For the purpose of this discussion, the stereophonic system is the section that is used for amplifying sounds—not the tuner (which receives signals from radio or TV stations).

The system consists of an amplifier, one or two loudspeakers, and a microphone. By talking and singing, sounds are emitted—notes characterized by different pitches (actually different wavelengths). The microphone picks up these sounds and notes (which are called input signals) and converts them to electrical signals. The amplifier then amplifies all the electrical signals and sends them into the loudspeaker, which converts them back to notes and sounds. They are reemitted by the loudspeaker, giving an output signal. If the system is sophisticated, the output signal from the loudspeaker reproduces the input signal fairly accurately as an amplified version. The system amplifies all the various frequencies of the

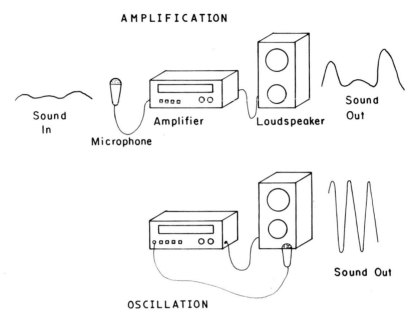

FIGURE 2.2   Feedback, gain, and oscillations—stereophonic system.

sounds in the input signal. If the input microphone is placed next to the output loudspeaker, the system emits a shrieking sound. This sound is a pure note of one frequency, independent of any input. The system starts to oscillate at one "resonance" frequency (where it has preferred amplification) instead of amplifying at many frequencies. An oscillator of this kind depends on two factors: (i) feedback between the input signal and the output signal (obtained by placing the microphone next to the loudspeaker) and (ii) amplification (obtained by use of an amplifier). To prevent oscillation, one can either stop the feedback (remove the microphone from the loudspeaker) or reduce the amplification (turn off the amplifier).

The laser similarly generates oscillations, but in the optical range rather than the acoustic range of frequencies. To obtain oscillation, both an amplifier and feedback system are needed. The idea of building a light amplifier, which is related to stimulated emission of radiation (hence the acronym laser) stems from the theoretical works of Einstein in 1916. This amplification (also called gain) appears in certain materials known as active media or lasing media. In order to obtain optical amplification, the lasing medium must be excited (pumped). Excitation is often provided by an electrical current. Under stimulated emission conditions, the intensity of the light that passes through the active medium is amplified synchronously.

Optical feedback is provided by a pair of mirrors which are located on both sides of the active medium. This is shown schematically in Fig. 2.3, where the light is reflected back and forth between two mirrors.

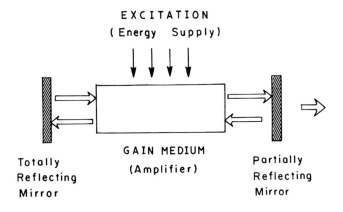

FIGURE 2.3 Optical oscillations in a laser.

The end result is similar to that in the case of the oscillations in the stereo system. The difference is that the oscillations from a stereo system give rise to acoustic waves, whereas the oscillations from a laser give rise to electromagnetic (EM) light waves. Optical oscillations or light of only one color is produced (much like the pure tone mentioned above) inside the lasing medium. This light can be extracted by slightly reducing the reflectivity of one mirror. One of the laser mirrors is not fully reflecting and transmits part of the light impinging on it. The light which emerges from this mirror is the narrow, collimated, monochromatic laser beam which is so useful in medicine and other fields.

The ruby laser will be used as an example. The first laser built and operated by Maiman in 1960, the ruby laser, was also one of the first lasers tested for medical applications. This laser is shown schematically in Fig. 2.4.

The active material in this case is a solid material, ruby, which is an inorganic crystal of aluminum oxide that contains a fraction of a percent of chromium impurities. The two end faces of the crystal are polished parallel to each other. Metallic layers are deposited on these faces to serve as laser mirrors. Whereas one of the mirrors reflects fully, the other reflects only partially. A helical flash lamp, similar to the flash lamp often used in photography, is used as a pumping (excitation) source. The ruby rod, with its mirrors, is inserted inside the flash lamp. When the flash is operated, it emits a burst of white light that excites the ruby rod. As a result, an intense pulse of deep red light (at a wavelength of 694 nm) is emitted from the ruby. This first flash of laser light was the first glimpse of modern optics and paved the way to the field of laser optics.

2.2.1.4 Classification of Lasers

A large number of lasers have been developed during the past three decades and may be classified according to the following characteristics:

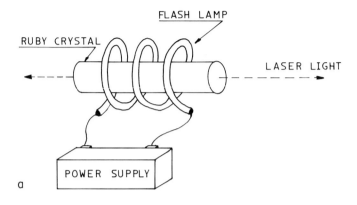

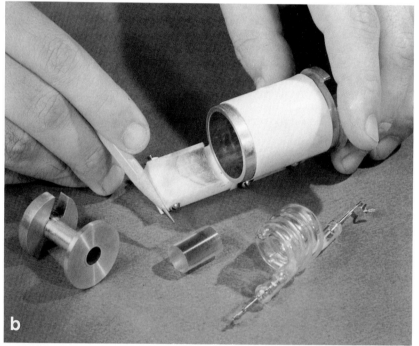

FIGURE 2.4 (a) Schematic drawing of the first laser. (b) Photograph of the first laser. (Courtesy of Dr. T. Maiman.)

(i) Lasing (active) medium: The lasing medium (where amplification takes place) can be a solid, liquid, or gas. Accordingly, the lasers are called solid-state lasers, liquid lasers, or gas lasers. Gas lasers such as CO_2 (carbon dioxide), excimer (e.g., XeF), and Ar (argon) are among the most important medical lasers. Dye lasers are examples of liquid lasers, and ruby or Nd:YAG (yttrium aluminum garnet doped with neodymium) lasers are solid-state lasers.

(ii) Wavelength: Lasers can emit radiation in the UV, visible, and IR parts of the electromagnetic spectrum. These are classified as UV lasers, visible lasers, and IR lasers.

(iii) Temporal behavior: Some lasers emit radiation continuously and are called continuous-wave (CW) lasers. Others emit bursts of radiation and are called pulsed lasers. Some of the pulsed lasers emit very short pulses (nanoseconds); others emit long pulses (milliseconds). The number of pulses emitted per second is called the pulse repetition rate. This rate can vary from very low (<1 pulse per second) to very high ($>10^9$ pulses per second) values. A few typical examples are shown in Fig. 2.5.

Research into the phenomena which led to the development and use of lasers is one of the culminations of human ingenuity. Some of the phenomena associated with lasers, however, also appear in nature. In the atmosphere of the planet Mars, for example, scientists have discovered stimulated emission in CO_2 molecules,

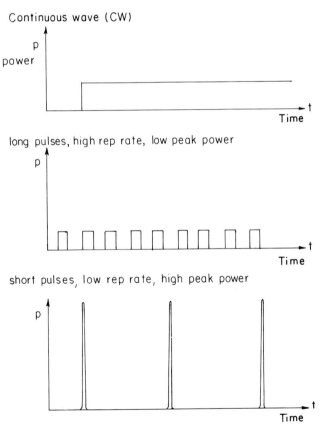

FIGURE 2.5 CW and pulsed laser beams.

which gives rise to optical gain. If Nature was truly interested in lasers, it would probably have added a pair of mirrors to the Martian atmosphere! This would have resulted in intense radiation at 10.6 μm—exactly like that emitted by a popular medical laser, the CO_2 laser.

2.2.2 Laser Physics—Principles

The interaction between tissue and lasers is governed by the properties of the laser beam as well as those of the tissue. Unfortunately, there is some confusion about the units to be used in describing all these properties. This confusion exists, in part, because the study of laser–tissue interactions is done by physicists, engineers, and physicians, all of whom use their own nomenclature. Beginning with CW lasers and continuing with pulsed lasers, this section endeavors to clarify the concepts that are related to laser light. Again, although these may not seem important for clinicians who "simply" want to treat patients, they are, in fact, crucial parameters that the clinician can vary in the commercial laser system. They determine the efficacy of the treatment and enable clinicians to compare their results with those obtained by others.

(i) Energy: Standard science textbooks define energy as the ability to do work. This naturally applies to the radiant energy in a laser beam. The energy units which are commonly used (see Section 10.1.2) are calories (cal) or joules (J), where 1 J = 0.24 cal.

(ii) Power: If energy E is emitted in time t, the average emitted radiant power P (also called flux) is $P = E/t$. The units commonly used are watts (W), where 1 watt is equal to 1 joule per second. Unfortunately, there is a difference regarding the specification of power for ordinary lamps and for lasers. The power specified in an incandescent lamp is the *input* power or the power consumed by the bulb (maybe because that is the power being paid for). For example, a 100-W lamp consumes 100 J/sec. The visible light energy emitted from this bulb is a small fraction of this total power (maybe 2%). In lasers, the *emitted* radiant power is generally specified. The laser also converts electrical power to light. When the laser emits 100 W, it may consume several thousand watts from the line, depending on its efficiency, which is usually low and seldom specified.

(iii) Power density: Lasers emit light in a parallel beam. The ratio of the emitted power (P) to the cross sectional area (A) is called the power density, or irradiance (I). The units of $I = P/A$ are W/cm^2. For example, lasers that emit 100 W in a beam of area 1 cm^2 have a power density of 100 W/cm^2. If a lens is inserted into the beam, the power does not change but the beam area may be reduced to 0.5 cm^2 and the power density doubles. The importance of this will be clarified when we consider the interaction of laser beams and materials.

(iv) Fluence: A laser beam may be operated intermittently, or the power delivered by the laser onto a given area may vary with time. The total energy deliv-

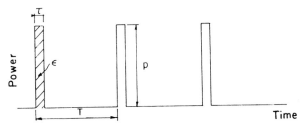

FIGURE 2.6 Pulsed laser beams: P = power (W); T = period (sec); $\varepsilon = P \cdot \tau$ energy per pulse (J); τ = pulse width (sec); $1/T = f$ repetition rate (sec^{-1}); average power $P_a = P \cdot \tau/T$.

ered, divided by the area (the energy per unit area) is called fluence. In a number of instances, fluence is the most important parameter for laser therapy.

These concepts can also apply to pulsed lasers beams. A pulsed laser emits a series of pulses as shown in Fig. 2.6. The duration of each pulse is denoted by τ and is measured in seconds. The number of pulses per second is called the repetition rate and is denoted by f. The time interval between two consecutive pulses is called the period T. It is clear that $T = 1/f$. Suppose that the amount of energy emitted from the laser for each pulse is ε joules. Through the duration of the pulse, the power is thus $P = \varepsilon/\tau$. This power is called the peak pulse power. If there are f pulses per second, the total energy emitted each second is εf, and for t_0 seconds, $E_0 = \varepsilon f t_0$. The average power P_{av} is obtained by dividing E_0 by the time t_0, so that

$$P_{av} = \varepsilon f t_0/t_0 = \varepsilon f = P\tau/T = \varepsilon/T.$$

EXAMPLE: A pulsed laser emitting $f_1 = 10$ pulses per second (i.e., 10 Hz), with an energy of $\varepsilon_1 = 0.1$ J $= 100$ mJ per pulse, emits an average power of $P_1 = \varepsilon_1 f_1 = 0.1 \times 10 = 1$ W (1 J/sec). If the energy per pulse is decreased to $\varepsilon_2 = 0.01$ J $= 10$ mJ and the repetition rate is $f_2 = f_1 = 10$ Hz, the average power is decreased to $P_2 = \varepsilon_2 f_2 = 0.01 \times 10 = 0.1$ W. If, however, the energy per pulse is $\varepsilon_3 = \varepsilon_2 = 0.01$ J $= 10$ mJ and the repetition rate of the same laser is increased to $f_3 = 100$ pulses per second (i.e., 100 Hz), the average emitted power will be increased to $P_3 = \varepsilon_3 f_3 = 0.01 \times 100 = 1$ W. For the trains of pulses in the first and the third case, the total energy emitted in 10 s will be 10 s $\times$ 1 W $= 10$ J. Although the first case illustrates a small number of high-energy pulses and the third case depicts a large number of low-energy pulses, the average power is the same for these two cases. The effects of these two exposures on materials or on tissue are usually not identical; that is, laser exposure effects are generally not simply proportional to energy.

2.2.3 Laser Physics—Advances

A special branch of optics that deals with the behavior of laser beams is sometimes called Gaussian beam optics (Wilson and Hawkes, 1987; Yariv, 1991; Siegman, 1986). Some of the important findings that are pertinent to our subject are discussed next.

2.2.3.1 Stimulated Emission and Amplification (Gain) in a Laser

As stated earlier, laser action depends on light amplification or gain in a medium. The basis for the gain in a medium depends on particular atomic or molecular species. In order to understand the origin of this gain, let us consider a simplified energy level diagram for such species, as shown in Fig. 2.7. There are a population N_1 of electrons in a lower energy level E_1 (sometimes called the ground state) and a population N_2 in an upper energy level E_2 (sometimes called the excited state).

An electron in the ground state E_1 can absorb a photon of energy $\Delta = E_2 - E_1$ and may be excited to the level E_2, as shown in Fig. 2.7a. This is an absorption

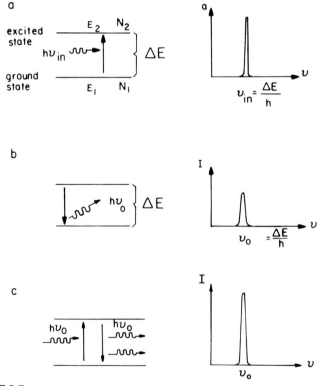

FIGURE 2.7 Atomic transitions: (a) absorption, (b) spontaneous emission, and (c) stimulated emission.

process. An electron in the upper level may return to the lower level with emission of a photon of an energy $\Delta E = E_2 - E_1$. There are two different emission processes:

(i) Spontaneous emission: The electron transition to the lower level is a random process. The emitted photon has the same energy (and wavelength) as the absorbed photon but is emitted at a random time and in a random direction. This is shown in Fig. 2.7b.

(ii) Stimulated emission: Emission occurs only in the presence of photons of energy $\Delta E = E_2 - E_1$. The electron transition is "triggered" by the absorbed photons and the emitted photons move in the same direction as the absorbed photons. The correlation in space and in time between the stimulating photons and the emitted photons is shown schematically in Fig. 2.7c.

Normally, the population N_1 in the lower state is much larger than the population N_2 in the upper state. When photons are incident on a collection of such atoms, they are absorbed. This leads to the classical absorption process. Under these circumstances, the spontaneous emission processes dominate. When a large number of atoms are considered, the radiation emitted from each of them will be random and the total radiation is called incoherent.

The situation is completely different if $N_2 > N_1$, a situation called population inversion. Einstein showed that in this case there must be stimulated emission. When we consider a large number of atoms for which there is population inversion, we must consider the presence of stimulating and emitted photons. All these photons have the same energy[2] (wavelength) and they move in the same direction. It turns out that they have the same phase and the same polarization.

Figure 2.7c shows schematically one input photon and two output photons of the same wavelength moving in the same direction. A similar process occurs when photons of suitable energy pass through a collection of atoms which are excited and exhibit population inversion. In this case, the number of emitted photons is larger than that of the incident photons. This is an amplification process and the medium is said to act as a light amplifier.

Several schemes promote the creation of a population inversion. Normally, a large amount of energy has to be supplied. This "pumping" energy may be in the form of direct electrical, light, or chemical energy. With sufficient pumping one can obtain light amplification, or gain. The gain medium may be gas, solid, or liquid.

As explained earlier, in a laser structure the medium that produces gain is placed between two mirrors that provide feedback. In addition to gain there are several loss mechanisms in the laser structure, such as scattering and absorption in the active medium and in the mirrors. The net gain is therefore the difference between the gain provided by the medium and the losses. Only if this net gain is positive does the structure emit light, or show lasing action.

---

[2] According to the Planck formula, there is a relation between the energy E of a photon and its wavelength λ: $E = hc/\lambda$, where h is Planck's constant and c is the velocity of light.

2.2.3.2 Transverse Electromagnetic Modes

As discussed earlier, a laser consists of a cavity and two mirrors. Excitation of the lasing medium in the cavity provides the imperative optical gain. The two mirrors supply the necessary optical feedback. The laser cavity is an optical resonator, in which light travels back and forth between the two mirrors. There are therefore electromagnetic fields which propagate back and forth in the cavity. Stable conditions inside the cavity are obtained only for a restricted number of combinations of EM fields, which are called *modes*. In particular, we are interested in transverse EM modes (TEM modes), in which both the electric and the magnetic fields are perpendicular to the beam axis. TEM modes are actually waves which travel in a direction that is close to the laser axis. After moving in a closed path between the two mirrors, they can replicate themselves. Different TEM modes are designated TEM_{pq}, where p and q are integers which specify the distribution of the nodes in the electromagnetic field. One may measure the intensity of the light inside the cavity, in particular the variation of the intensity with its distance from the beam axis. This intensity distribution is different for the various modes TEM_{pq}. The same intensity distribution appears outside the cavity, in the beam emitted from the laser. A few typical distributions for various TEM modes are shown schematically in Fig. 2.8.

Intensity distribution is not just a mathematical curiosity; it is extremely important for laser–tissue interaction and in particular for laser surgery. In order to obtain a sufficient power density to vaporize the tissue, without affecting adjacent tissue, the laser surgeon attempts to focus the laser beam onto the smallest possible spot. For example, Fig. 2.8e illustrates four partial beams. When trying to focus the beam, four spots are obtained. The power density is rather low and tissue removal may be hindered. The same is true for the doughnut-shaped beam in Fig. 2.8d. Although the total power, as measured by a power meter, may be high in such a beam, the power density at the focal spot will be relatively low. In order to obtain efficient focusing and high power densities, it is generally advantageous to obtain the mode designated TEM_{00}, as shown in Fig. 2.8a.

2.2.3.3 Gaussian Beam

The TEM_{00} mode is the desirable mode for most applications of lasers. The electric field variation is radially symmetric and is given by the formula

$$E(r) = E(0) \exp(-r^2/w^2).$$

In this formula, r represents the radial distance from the beam axis and w is a parameter defined as the beam radius. This formula is called a Gaussian distribution, after the great mathematician Gauss who developed this expression.[3]

The intensity I is proportional to E^2. The intensity distribution for this mode is therefore given by the formula

$$I(r) = I(0) \exp(-2r^2/w^2). \tag{2.1}$$

---

[3] The Gaussian distribution has the unique property that it remains unchanged even after the beam is reflected back and forth between the mirrors in the laser cavity.

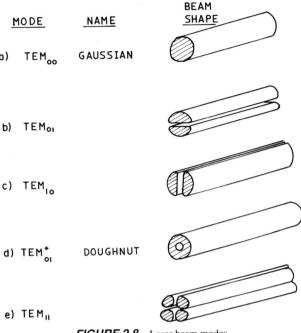

FIGURE 2.8   Laser beam modes.

Such a beam is called a Gaussian beam and the plot of $I(r)$ is shown schematically in Fig. 2.9. At a radial distance $r = w$, the intensity is $I(0)/e^2$ or approximately $I(0)/10$.

The power included within a radius r_A is obtained by integrating $I(r)$ between 0 and r_A. Integration between 0 and ∞ gives the total power in the beam, P. It can be shown that the intensity $I(0)$ at the beam center is given by

$$I(0) = 2P/(\pi w^2). \tag{2.2}$$

It follows that about 90% of the total power P is contained in the spot whose radius is w (or whose diameter is $2w$). The quantity $D = 2w$ is defined as the beam diameter, as shown in Fig. 2.9.

Let us now suppose that there is a series of *apertures* of various diameters d through which we try to transmit the beam. For an aperture of diameter $d = D$, it is possible to transmit $\approx 0.9P$ (i.e., roughly 90% of the power). For smaller apertures, a smaller fraction of P, according to the preceding formula, is transmitted. This is also shown in the Fig. 2.9. The importance of these numbers is clarified when we discuss optical fibers.

2.2.3.4 Beam Divergence

As the laser beam propagates, its diameter increases. This phenomenon is called beam divergence (see Fig. 2.10). Normally, there is a minimal spot (called

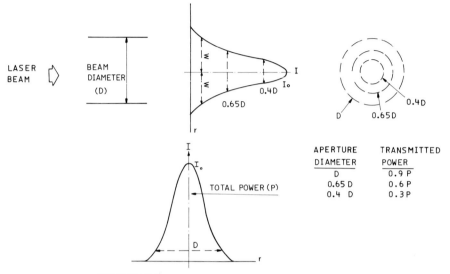

FIGURE 2.9   Gaussian laser beam: intensity distribution.

the beam waist) whose radius is w_0 and which is often situated inside the laser (or outside, very close to the output window). If z is the distance in a horizontal direction from the position where the minimum spot w_0 occurs, then the radius $w(z)$ of the beam at this value of z may be calculated (Siegman, 1986). This is actually the value which should be used in Eq. (2.1). It was found that for large distances from the output mirror, in a region called the far field, $w(z)$ is proportional to z:

$$w(z) \approx (\lambda z)/(\pi w_0) \tag{2.3a}$$

$$2w(z) \approx (2/\pi)(\lambda z/w_0). \tag{2.3b}$$

It can be shown that for the diverging beam Θ, half the angle of divergence, is given by the equation

$$\Theta = (2/\pi)(\lambda/2w_0). \tag{2.4}$$

This is shown schematically in Fig. 2.10, where Θ is measured in radians and

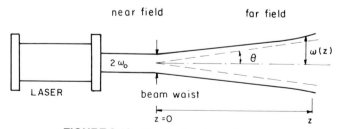

FIGURE 2.10   Gaussian laser beam: divergence.

the *full* divergence angle is 2Θ. We can approximate $4 \approx \pi$ and thus write the equation

$$2\Theta = (4/\pi)(\lambda/2w_0) \approx (\lambda/2w_0). \qquad (2.5)$$

This is similar to the classical formula describing the divergence of light through an aperture of diameter $2w_0$.

It should be pointed out that the beam divergence is orders of magnitude lower than the divergence that one observes in ordinary (incoherent) light sources. This is a result of interference effects.

EXAMPLE: An HeNe laser typically has a minimum beam diameter of $2w_0 = 1$ mm. The wavelength of the HeNe laser is $\lambda = 0.63$ μm $= 0.63 \times 10^{-3}$ mm. The full divergence angle calculated by Eq. (2.5) is roughly 10^{-3} radians $= 1$ milliradian (i.e., $0.06°$). At a distance of 10 m from the laser, the beam diameter will thus be only 1 cm. The divergence may be further decreased by sending the beam through a special telescope. This is the property which made it possible to send laser beams to the moon.

2.2.3.5 Beam Focusing

Some of the most important applications of lasers involve focusing the beam on a tiny spot with a lens. If a lens of focal length f is inserted in the path of a laser beam, the Gaussian beam is focused. Let us assume that the z coordinate of the focal spot is $z = 0$ and that of the lens is $z = -f$. The beam radius at the focal spot is $w_{(z=0)} = w_0$. Assuming that the beam exactly fills the lens, the diameter at the lens is $2w_{(z=-f)} = D$, where D is the lens diameter, as shown in Fig. 2.11. It is interesting to note that the equations which describe Gaussian beam divergence are valid also for $z < 0$. In particular, Eq. (2.3b) may now be written as

$$2w_{(z=-f)} = D = (2/\pi)(\lambda f/w_0) \qquad (2.6)$$
$$w_0 = \lambda f/(\pi w_{(z=-f)}) = (2/\pi)(\lambda f/D).$$

In optics (and in photography) it is common to describe lenses by their f/D value. This quotient is called the f-number or $f/\#$. We could therefore write an equivalent expression:

$$f/D = f/\#.$$

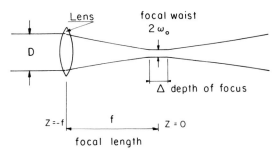

FIGURE 2.11   Gaussian laser beam: focusing.

By substituting this definition in (2.6), the diameter of the focal spot is obtained by

$$2w_0 = (4/\pi)(\lambda f/D) \approx \lambda(f/\#). \qquad (2.7)$$

EXAMPLE: A laser beam with a wavelength of λ is focused with a lens of diameter D and a focal length f. In the case $D = f$ (i.e., $f/\# = 1$) where the diameter of the lens is equal to the focal length, the beam is focused to an extremely small spot whose diameter is equal to the wavelength λ.

Lenses generally have $f/\#$ values which are greater than 1. Yet even then, the laser beam focuses to a remarkably small spot, measuring a few wavelengths. The power density in such a spot is therefore extremely high.

As one moves away from the focal spot, the diameter of the beam increases. The distance (in the normal direction) over which the spot diameter does not increase by more than 5% is called the depth of focus and is approximately

$$\Delta = (8\lambda/\pi)(f/\#)^2 \approx 2\lambda(f/\#)^2. \qquad (2.8)$$

EXAMPLE: An HeNe laser beam ($\lambda = 632.8$ nm $= 0.6328$ μm) of diameter 1 mm is focused by a lens with the same diameter D and a focal length $f = 10$ mm. The f-number is $f/\# = 10$. The focal spot diameter, as given by Eq. (2.6), is $2w_0 \approx (0.63)(10/1) \approx 6$ μm. The depth of focus is $\Delta \approx 2(0.6328$ μm$)(10)^2 \approx 130$ μm. If the focal length of the lens is increased by a factor of 10 to $f = 100$ mm, then $f/\# = 100$. In this case, the focal spot diameter is also increased by 10 and the depth of focus by 100!

The variation of the spot size and the depth of focus with these lens parameters are shown in Fig. 2.12. By using a lens with a short focal length, one obtains a small spot size in addition to a small depth of focus. In laser surgery, the physician must focus the beam to a small spot. Although a short focal length may seem advantageous, there is little room for error. If the surgeon applies the beam slightly out of focus, the spot size increases, the power density decreases, and the ability to cut decreases rapidly. With a longer focal length, the spot size is larger, but the depth of focus also increases. Although one loses some power density, the operator gains much more flexibility because he or she may move the hand piece up or down and still remain in focus.

2.3 MEDICAL LASERS

A brief explanation of the operating characteristics of the lasers that are commonly used in medicine illustrates the issues raised in the preceding section. The most important lasers and their emission wavelengths[4] are shown schematically in Fig. 2.13. Note that the wavelength scale is logarithmic and not linear.

---

[4] The wavelength λ may be measured in units of either micrometers (μm) $= 10^{-6}$ m or nanometers (nm) $= 10^{-9}$ m. Both units are used interchangeably in this book.

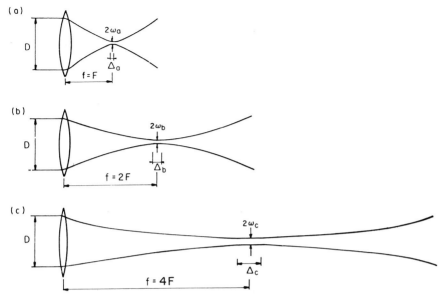

FIGURE 2.12 Gaussian laser beam: focusing by different lenses (spot size and depth of focus).

2.3.1 Medical Lasers—Fundamentals

In this section, two gas lasers are described: the CO_2 and the Ar ion laser, and one solid-state laser, the Nd:YAG laser (Winburn, 1987). These lasers have different characteristics and different roles in medicine. Because the CO_2 laser radiation is highly absorbed in tissue, it is used mostly for cutting. In contrast, the Nd:YAG laser radiation is only weakly absorbed in tissue and thus is used mostly for heating large volumes of tissue.

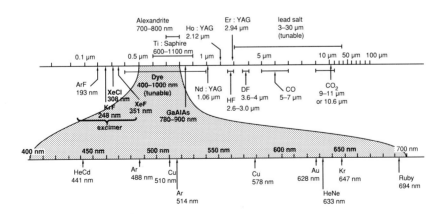

FIGURE 2.13 Wavelengths of popular lasers.

2.3.1.1 Carbon Dioxide Laser

The CO_2 gas laser, one of the first lasers developed in the early 1960s, is still one of the best for surgery. Most gas lasers consist of a tube, much like the glass tube of a fluorescent lamp. The gas is either enclosed in this tube or contained in a gas bottle and pumped through the laser tube during operation. Excitation is provided by an electric discharge through the gas in the laser tube. The main difference between this setup and a regular fluorescence lamp is the presence of two mirrors at each end of the tube that provide feedback. In the CO_2 laser, the lasing medium is actually a mixture of CO_2, nitrogen, and helium. The laser emission is in the midinfrared part of the spectrum at $\lambda = 10.6$ μm.

In most commercially available medical laser systems, the beam is directed toward the operating site via a set of mirrors called an "articulated arm" (see Section 2.3.4), at the end of which is a focusing lens that focuses the laser beam to a tiny spot. The laser energy that is absorbed in the tissue under this spot heats the tissue to a high temperature and evaporates it. In order to perform laser surgery, at least 10–20 W of CO_2 laser energy is needed. Some of the lasers used may emit up to 100 W either continuously or in pulses.

From the practical point of view, the CO_2 laser is a well-developed and reliable system. In order to operate a typical axial flow laser, a few standard items may be required, a special electrical supply, a special water outlet for cooling water for the laser, and a supply of CO_2 gas bottles. Thousands of hospitals and laboratories worldwide use these lasers. Small sealed-off CO_2 lasers are less technologically demanding.

Persons unfamiliar with the CO_2 lasers should be aware that this laser emission is in the infrared and is therefore not visible to the naked eye. In commercial lasers, a low-intensity HeNe laser aiming beam is sent parallel to the CO_2 laser beam. It indicates the path of the invisible beam and thus provides vital assistance for beam placement. Since the power levels used in surgery are rather high, the laser should be used with caution. If the high-intensity infrared beam is pointed at a reflecting object, a small fraction of the beam may be reflected toward the operator, where it may cause harm. The organ most sensitive to this radiation is the eye, where a reflected CO_2 laser beam is highly absorbed by the cornea and may cause severe damage. In order to protect the eyes from the beam, eyeglasses with lenses made of plain glass or plastic must be worn. Plastic and plain glass are transparent to visible light but totally block the CO_2 laser IR beam. Apart from this risk, however, the CO_2 is one of the safest lasers.

2.3.1.2 Nd:YAG Laser

This laser is quite similar to the ruby laser that was described earlier. The Nd:YAG laser is a solid-state laser whose active medium is a solid rod of the crystal yttrium aluminum garnet (YAG). The YAG, an artificial, diamondlike structure, is not pure but includes impurity ions of the rare earth element neodymium (Nd). The emitted radiation is not in the red (like that of the ruby laser) but rather in the near infrared—1.06 μm. Lasers with power levels lower than 60 W can be operated without water cooling, using a single-phase 220-V (or 110-V) outlet.

The Nd:YAG laser radiation is often used for heating, coagulation, or treatment of tumors, where use is made of its deep penetration in tissue. Because the necessary power levels for these applications have been found to be tens of watts, many of the Nd:YAG lasers operate in this power range. For power levels above 60 W, a special electricity supply and cooling water for the laser head are sometimes needed. Since the laser emission in the near infrared is also invisible, the laser must be operated with caution. Moreover, because the emission at 1.06 μm is not absorbed by the cornea, as in the case of CO_2 laser, it may penetrate the eyes and cause retinal damage. When operating the laser, special goggles must be worn to block out the laser near-IR emission while transmitting visible light.

2.3.1.3 Argon Ion Laser

The Ar ion laser utilizes ionized argon gas as its lasing medium and its structure is somewhat different from that of the CO_2 laser. Since the Ar ion laser is of low efficiency in terms of converting electrical power to laser power, it requires a large current and therefore a large power supply. Water cooling is usually needed.

The Ar ion laser is used mostly in ophthalmology. The laser beam, which may be focused on the retina, is used to prevent retinal detachment in addition to other medical and surgical applications. For all these applications, a few watts to tens of watts of laser output are necessary.

The Ar ion laser emission is in the blue–green part of the spectrum. This emission is well transmitted and well focused by the eye and may cause damage to the retina. When using the Ar laser, special goggles which block out the blue–green laser emission must be worn to protect the eyes.

2.3.2 Medical Lasers—Principles

The three lasers mentioned in Section 2.3.1 (CO_2, Nd:YAG, and Ar ion lasers) are the ones most commonly used for medical applications. The following section provides a more detailed explanation of these lasers in addition to other useful medical lasers (Meyers, 1991; Hecht, 1992). Lasers are listed by wavelength, starting from the IR and ending in the deep UV. Their properties are summarized in Tables 2.2 and 2.3. A wealth of information on commercial lasers may also be obtained from various "buyers' guides" that are updated annually and are available from various journals (e.g., *Laser Focus World*, *Lasers and Photonics*).

2.3.2.1 Carbon Dioxide Laser

The CO_2 gas laser is very popular for several reasons. For one, the laser wavelength at $\lambda \approx 10$ μm is highly absorbed in tissue and seems to be most suitable for tissue removal. Other reasons are more technical, such as size, cost, and reliability. Furthermore, the CO_2 laser is highly efficient. Overall efficiency is measured by the ratio of emitted power (laser emission) to the input power (electricity)

TABLE 2.2 Continuous (CW) Laser Properties

Laser (general)	Wavelength	Laser medium	Maximum power
HeCd	325.0 nm	Gas	50 mW
	442.0 nm		150 mW
Ar ion	488.0 nm	Gas	20 W
	514.5 nm		
Kr ion	413.1 nm	Gas	5 W
	530.9 nm		
	647.1 nm		
Dye (Ar pumped)	400–1000 nm	Liquid	2 W
HeNe	632.8 nm	Gas	50 mW
GaAlAs	750–900 nm	Semiconductor, single laser, array of lasers	100 mW 10 W
Nd:YAG	1060 nm (1.06 μm)	Solid state	600 W
HF	2600–3000 nm (2.6–3.0 μm)	Chemical	150 W
CO_2	10,600 nm (10.6 μm)	Gas	100 W

and for this laser is about 10–20%. Thus, in order to operate a laser which emits 100 W, less than 1000 W (not more than an electric heater) need be supplied. Most other types of lasers have an overall efficiency of about 1–2%. For an output of 100 W, these lasers require many kilowatts of input power, necessitating the use of special electric power lines and additional expenses. The gases used in the CO_2 laser are nontoxic, noncorrosive, and inexpensive. The only expensive component is helium (He) which can be reprocessed. In most commercially available CO_2 lasers, the gas mixture of CO_2, N_2, and He flows through the laser at a low rate. These are called flowing gas lasers and are shown schematically in Fig. 2.14. The gas mixture is available premixed in bottles. Within the system, there is a pump which pumps the gases through the laser cavity. In practical terms, a constant supply of gas bottles must be readily available in the operating room. Other CO_2 lasers have no flowing gas but the gas mixture is contained in a sealed tube, as shown schematically in the same figure. Even when sealed, the gases need to be replenished regularly.

CO_2 lasers may also be classified according to the excitation process of the lasing action. The lasing action usually occurs through a direct current (DC) through the gas. Some lasers are operated by long (milliseconds) electric pulses (pulse-excited lasers) or by radio-frequency (RF) currents (RF-excited lasers). Excitation may also occur by transverse electrodes, which enable the laser to be operated at atmospheric pressure. These lasers are called TEA lasers (transversely excited atmospheric pressure). An additional way to classify CO_2 lasers, which applies to other lasers as well, is by their temporal behavior. Some lasers emit radiation continuously (CW). Others emit radiation with fairly long pulses (milli-

TABLE 2.3 Pulsed Laser Properties

Laser (general)	Wavelength (nm)	Laser medium	Pulse duration	Maximum energy (J per pulse)	Maximum repetition rate (Hz)	Maximum average power (W)
Excimer	193	ArF gas	5–25 nsec	0.5	1000	50
	249	KrF gas	2–50 nsec	1	500	100
	308	XeCl gas	1–300 nsec	1.5	500	150
	351	XeF gas	1–30 nsec	0.5	500	30
Dye: Ar pumped	400–1000	Liquid dye	3–50 nsec	0.1	100	10
flash pumped	350–1000	Liquid dye	0.2–30 μsec	50	50	50
Metal vapor	628	Au vapor	15 nsec	10^{-4}	10^4	1
Semiconductor	750–1550	Single junction	0.2–2 μsec	10^{-4}	10^4	0.1
(AlGaAs)		Array of lasers	2–200 μsec		100	1
Nd:YAG	1.06 μm	solid	10 nsec	1	20	20
Er:YAG	2.94 μm	solid	200 μsec	2	20	30
HF	2.6–3 μm	chemical	50–200 nsec	1	20	3
CO_2 pulsed	10.6 μm	gas	10 μsec–10 msec	5	1000	100
waveguide		gas	1 μsec	0.01	5000	50
TEA		gas	20 nsec–10 μsec	150	1000	30

SEALED

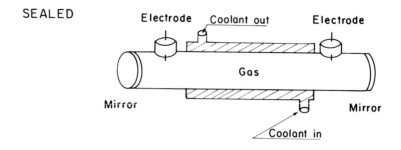

FLOW

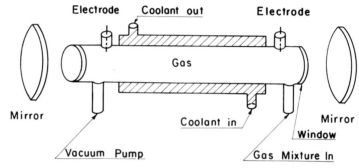

FIGURE 2.14 Schematic drawing of a gas laser.

second) or with short pulses (microsecond). Some can be operated in several temporal modes.

A few of the common CO_2 laser systems are described below from a practical point of view:

(i) CW lasers: Most of these lasers are gas flow lasers and are DC excited. The power output of the lasers depends on their length; one may obtain about 100 W from a laser with a length of 1 m. The typical beam diameter is 5–10 mm. These lasers weigh more than 100 kg, and need to be wheeled from place to place. Their operation requires water cooling and a constant gas supply. These are the lasers most commonly used in hospitals for laser surgery. Other CW lasers include the sealed-tube type. The maximum power output of these lasers is about 100–200 W.

(ii) Pulsed lasers: Although they are similar in their construction and properties to the CW lasers, pulsed lasers operate in a pulsed mode. Laser pulses of few milliseconds (10^{-3} sec) can be obtained giving a few hundred millijoules per pulse and a repetition rate of a few hundred pulses per second. The average power may reach 50—100 W.

(iii) Short pulsed lasers: These are the TEA lasers, mentioned above, which are fairly large and require water cooling. With an energy of tens to hundreds of millijoules per pulse and repetition rates of up to a few hundred pulses per second, the pulses may last tens of nanoseconds (10^{-9} sec) or a few microseconds (10^{-6} sec).

(iv) RF-excited lasers: These are operated by an RF power supply and emit a continuous train of short pulses at a rate of about 10,000 pulses/sec. The average power output of such lasers may reach about 250 W.

(v) Waveguide lasers: This is one type of RF-excited laser in which the lasing takes place in a very thin ceramic tube, which is actually a hollow waveguide. Lasers of this type are sealed and are distinguishable by their compactness. They weigh only a few kilograms and require no gas supply. They can continuously emit an average power of more than 100 W with water cooling and lower average power (e.g., 10 W) without cooling. The typical beam diameter is 1–2 mm. These waveguide lasers are small enough to be hand held and the emitted power is sufficient for laser surgery.

2.3.2.2 Er:YAG and Ho:YAG Lasers

The Er:YAG laser is a solid-state laser which is quite similar to the Nd:YAG laser, except that the impurity in the YAG crystal is the rare earth erbium (Er) rather than neodymium (Nd). Because emission is at 2.94 μm, the peak of the absorption curve of water, a major component of tissue (see Fig. 3.3), this laser is potentially very useful in surgery.

Holmium (Ho) is also used as an impurity to form the Ho:YAG laser. It has been found that this dopant by itself is not sufficient to form a practical laser. Two other dopants, such as thulium (Tm) and chromium (Cr), must be added to serve as sensitizers. The Ho:YAG laser has an emission wavelength of 2.1 μm. This laser is potentially useful for medical applications because this wavelength is attenuated by tissue, but at the same time it can easily be transmitted through fused silica fibers (see Chapter 4).

Both the Er:YAG and the Ho:YAG lasers operate in the pulsed mode with pulses of about 200 μsec and several hundred millijoules per pulse. The repetition rate is a few pulses per second. Each of the 200-μsec pulses actually consists of a train of about 20 shorter pulses (typically 1 μsec long). Both lasers can be pumped by semiconductor lasers.

There are several other solid-state lasers that are based on doped crystals, such as YAG (yttrium aluminum garnet) or YLF (yttrium lithium fluoride), and some are available commercially.

2.3.2.3 Neodymium Yttrium Aluminum Garnet (Nd:YAG) Laser

This is a solid-state laser, whose structure is shown schematically in Fig. 2.15. The active medium is a crystalline material, in this case a synthetic YAG crystal doped with roughly 1–2% of the rare earth Nd. The main emission wavelength of the Nd:YAG laser is 1.064 μm in the near IR. The Nd:YAG crystal is excited (pumped) by flash lamps. The total efficiency of the laser is very low, about 0.1–1.0%. Heat is generated during the operation of the laser, necessitating water cooling. Also, in order to obtain a power output of 100 W, for example, an input

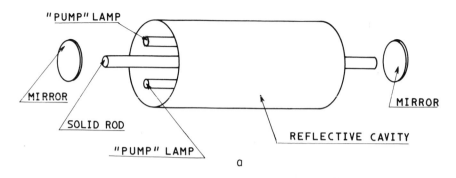

FIGURE 2.15 (a) Schematic drawing of a solid-state laser. (b) Photograph of a solid-state laser. (Courtesy of Schwarz Electro Optics.)

electrical power of about 7–10 kW is necessary. This requires a special line, of perhaps 220 V, three phases, and 40 A. Water cooling or a special power supply may not be necessary if the CW power output is less than 50 W. One of the advantages of an Nd:YAG laser is that it does not require the constant supply of consumable gases. It is fairly rugged and has a variety of applications. The Nd:YAG laser can be operated either continuously or in the pulsed mode, with pulse durations ranging from picoseconds (10^{-12} sec) to milliseconds.

(i) CW Nd:YAG lasers: These lasers are excited by continuous lamps, such as quartz halogen or arc lamps. Typically, in such an operation, about 100 W output can be obtained, and industrial-type lasers may reach 1000 W.

(ii) Pulsed Nd:YAG lasers: These lasers are excited by pulsed (e.g., flash) lamps, at repetition rates up to 200 Hz. The typical laser pulses are 0.1–10 msec in duration, with energies of 0.1–100 J per pulse; the average power may reach a few hundred watts.

(iii) Q-switched Nd:YAG laser: In this mode, the excitation is carried out with CW or with pulsed lamps. An additional pulsing method, called Q switching (see Section 2.3.3.5), however, generates very short laser pulses (10–30 nsec). A typical Q switched laser operates at tens of pulses per second with a pulse duration of tens of nanoseconds and energies measuring tens of millijoules per pulse.

(iv) Diode-pumped Nd:YAG laser: The laser light emitted from semiconductor lasers can be used to excite (pump) the Nd:YAG crystal. Actually, a phased array of GaAs:GaAlAs lasers (see Section 2.3.3.3) is needed to excite the crystal. The semiconductor lasers are much more rugged than the flash lamps used in ordinary Nd:YAG lasers, and the pumping efficiency is higher. As a result, the total efficiency of this type of laser is 5–8%, which is much higher than that of an ordinary Nd:YAG laser. Diode-pumped Nd:YAG lasers are still in the early stages of development. At present the CW power output is less than 1 W, but this is likely to increase in the future.

2.3.2.4 Dye Lasers

A liquid dye, such as rhodamine 6G, composes the active medium in a dye laser. The dye is optically excited (i.e., pumped) by a light source, which may be a lamp or another laser. The dye lasers operate continuously or in the pulsed mode. CW dye lasers are pumped by other CW lasers (e.g., ion lasers) and pulsed dye lasers are pumped by flash lamps or by pulsed lasers (e.g., excimer). The power output of these lasers is generally several watts. Their emission covers a range of wavelengths (which depends on the specific dye) in the visible part of the spectrum. Their main advantage is that the emission wavelength can be changed ("tuned"). Some of the tuning ranges are as follows: stilbene, 410–470 nm; coumarin, 450–500 nm; rhodamine, 550–650 nm. As a system, these dye lasers are often complex and difficult to operate.

There are applications which make use of the accurate tuning. For example, a chromophore (coloring agent) in tissue has strong absorption in a narrow spectral region. In order to facilitate such absorption, the dye laser must be tuned to this particular spectral range. The same applies for excitation of characteristic fluorescence from tissue at a specific wavelength.

2.3.2.5 Tunable Solid-State Lasers: Alexandrite and Ti:Sapphire

A few solid-state lasers have a tunable output wavelength. These lasers are somewhat similar in construction to the Nd:YAG laser. They may be pumped optically by a flash lamp, by another laser, or even by a semiconductor laser (much like Nd:YAG), and they are operated either continuously or in the pulsed mode. The main difference is that a special tuning element is incorporated in the structure. This is an optical element that makes it possible to select a narrow wavelength range in which the laser emits within a broader range in which the laser can operate. The tunable solid-state lasers have not yet been used widely in medicine but they are potentially very useful. In this section we discuss two of these lasers.

The alexandrite laser is based on the doped crystal $Cr:BeAl_2O_4$ and it has a tuning range of 720–800 nm. An alexandrite laser is typically operated in the pulsed mode, at tens of pulses per second with pulse lengths of 200–1000 nsec and tens of millijoules per pulse.

The Ti:sapphire laser is based on the doped crystal $Ti:Al_2O_3$ and it has a tuning range of 670–1100 nm. The Ti:sapphire laser is often operated in the pulse mode, with tens of pulses per second, pulse lengths of few microseconds, and energies up to hundreds of millijoules per pulse. It may also be operated continuously (CW) with a power output of several watts.

2.3.2.6 Semiconductor Lasers

These lasers are very different from all the lasers mentioned above. The typical size of a semiconductor laser is equal to that of a grain of salt, and the laser is shown schematically in Fig. 2.16. The active material is a semiconducting crystal, usually based on GaAs or similar compounds such as GaAlAs, InGaAs, or AlGaInP. A current that passes through a "sandwich," two types of layers consisting of such compounds (called p-type and n-type), gives rise to an emission in the near IR (0.8–1.5 μm). This emission is very similar to the visible emission of light-emitting diodes (LEDs), which are common in various instruments. When two mirrors are added to the structure, it operates like a tiny laser (instead of an LED).

These semiconductor lasers can be operated in the CW mode or in the pulsed and even short pulse modes. The emission wavelength of such a laser depends on the compound used. The overall efficiency of semiconductor lasers is more than 30% and is among the highest. They can therefore be operated without special cooling, at relatively low electric currents. Promising developments have resulted in continuous power output of several tens to hundreds of milliwatts from such a tiny laser.

2.3.2.7 Helium–Neon (HeNe) Laser

This gas laser is based on a mixture of helium and neon. Normally, the laser is sealed and operates in a quasicontinuous mode with an average power of a few milliwatts. The emission is in the red part of the spectrum at 632 nm. The laser is portable, lightweight, and very reliable. It is used mostly as an aiming beam in IR (e.g., CO_2) laser systems. In these cases, the laser beam is invisible and an HeNe laser beam is used jointly with the IR beam. If it is necessary to move the IR laser beam, as in the case of an articulating arm, or to focus on a fiber, the HeNe laser beam helps to aim the IR beam.

2.3.2.8 Metal Vapor Lasers

The active medium consists of a mixture of the gas Ne with the vapor of a metal, such as copper or gold. The laser operates at rather high temperatures (about 1500° C), and excitation is provided by electric discharge. The emission wavelength depends on the metal used. For copper the emission is in two green lines at 510.6 and 578.2 nm, while gold emits one red line at 628 nm and a weak UV line at 312 nm. Metal vapor lasers are pulsed lasers with pulse repetition rates of about 5–10 kHz. The average power of the copper vapor laser may reach a few tens of watts and that of the gold vapor laser only a few watts. One of the applications of the gold vapor laser is in photochemotherapy cancer treatment (see Section 3.7.2), in which relatively high power at 630 nm is necessary.

The metal vapor lasers have an overall efficiency of less than 1%. They require a few kilowatts of electrical input, as well as water cooling of the laser head. A major problem associated with these lasers is that the metal vapors must be kept

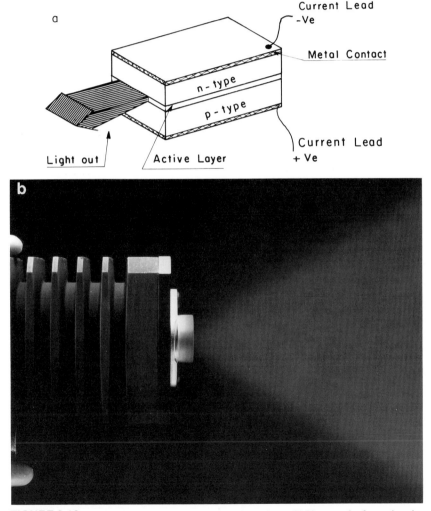

FIGURE 2.16   (a) Schematic drawing of a semiconductor laser. (b) Photograph of a semiconductor laser. (Courtesy of Spectra Diode Laboratories.)

at a high temperature in the laser tube. This difficult technical problem has not yet been solved satisfactorily and many of these lasers suffer from reliability and maintenance problems.

2.3.2.9 Argon (Ar) and Krypton (Kr) Ion Lasers

Argon (Ar) gas forms the basis of the active medium in the Ar laser and, unlike CO_2 in the CO_2 laser, the gas must be ionized. The laser is thus called an argon ion laser, or simply an Ar laser. The emission of this gas is in the UV and visible part of the spectrum, with two prominent emission lines at 488.0 nm (blue) and 514.5 nm (green). One of the first lasers to be developed, it is widely used in ophthalmology. The visible laser beam is well transmitted by the eye and can be readily focused on the retina, as required for detached retina treatment. The Ar laser works in the CW mode, with power output of several watts in the two prominent lines. Its total output may reach approximately 20 W. Since the efficiency of this laser is fairly low, it requires a relatively high current and water cooling. There are new Ar lasers with lower CW power (10 W) that are air cooled.

The Kr laser is similar to the Ar laser, but it is based on ionized krypton gas. This laser has numerous emission lines in the UV and visible spectral range. Some of the prominent lines are at 413.1, 530.9, and 752.5 nm. The strongest line is at 647.1 nm (red). The efficiency of the Kr laser (and the power output) is lower than that of the Ar laser. The power output is normally less than 1 W in each of the prominent lines.

2.3.2.10 Helium–Cadmium (HeCd) Lasers

This laser is similar in structure to the HeNe laser but the gas used is a mixture of helium (He) and ionized cadmium (Cd). In principle, this is a metal vapor laser. The HeCd laser operates continuously (CW) with emission in several lines in the UV and the visible range of the spectrum. It has two prominent lines at 441.6 nm (blue) and 325 nm (UV). The overall efficiency of the laser is fairly low (less than 0.05%) and total power output may reach about 50 mW for the UV line and 150 mW for the blue line.

2.3.2.11 Excimer Lasers

These lasers are very similar in construction to the CO_2 TEA gas lasers, but the gases used are rare gas halides such as ArF, KrF, XeCl, or XeF. These diatomic molecules (i.e., dimers) are stable only in their excited state, not in their ground state. The name excimer is derived from excited dimers. The emission from each excimer laser varies for the different gases (see Table 2.3). The emission of all the excimer lasers is in the UV at wavelengths shorter than 350 nm. Because absorption in tissue is high at all UV wavelengths, these lasers are potentially useful in surgery. They always operate in the pulsed mode with short pulses (10–100 ns). Pulse energies measure up to a joule at a repetition rate of up to 300 Hz and average power output is up to tens of watts. The overall efficiency of the lasers is less than 2%. There are several problems associated with these lasers. The gases used are both corrosive and toxic. Precautions have to be taken to prevent leakage of gases, which may become a health hazard. The lasers need technical service every few months. Questions have been raised regarding the carcinogenicity of the UV emission. In addition, the beam is rather nonuniform and may contain "hot spots" (i.e., areas of high intensity). If solutions to all these problems are found, the lasers will be very useful for a variety of applications.

2.3.3 Medical Lasers—Advances

Numerous lasers may turn out to be useful in medicine in the future. Descriptions of these lasers can be found in texts mentioned in the references at the end of the chapter. There are several lasers which, although potentially useful for medical applications, have not yet been widely tested. In this section we discuss a few of these lasers.

2.3.3.1 Carbon Monoxide (CO) Laser

This laser is similar in some respects to the CO_2 laser, except that the gas used is carbon monoxide. The commercially available CO lasers are modified versions of the CO_2 lasers. Yet there are some significant differences. Many of these lasers are operated at low temperatures (e.g., cooled with dry ice powder). This adds to the complexity of the system. Also, the gases used are toxic. This presents problems, especially if we are talking about a gas flow system where leaks may occur. The CO laser typically operates in the CW mode, with several emission lines between 5 and 6 µm and a power output of several watts. This laser is interesting because the emission in the mid-IR is highly absorbed in tissue and thus good for surgical applications. In addition, it may be easier to transmit the $\lambda \approx 5$ µm laser beam through optical fibers than the $\lambda = 10.6$ µm emission of CO_2 lasers, as we will see in Section 4.7.

2.3.3.2 Hydrogen Fluoride (HF) Chemical Lasers

If the energy needed for lasing is generated by a chemical reaction, the lasers are called chemical lasers. Hydrogen fluoride (HF) lasers are gas lasers that are based on the reaction between free fluorine and hydrogen that produces excited HF molecules in an exothermic reaction. The main attraction of the HF laser is its ability to generate extremely high power, which is important for the military. In addition, the laser emission consists of many discrete lines in the wavelength range 2.6–3.0 µm, including wavelengths around 2.9 µm that are highly absorbed by water. This characteristic makes the HF laser interesting for surgical applications as well.

Commercially available pulsed HF lasers are similar in appearance to CO_2 TEA lasers or excimer lasers. Because the total efficiency of the lasers is 0.1–1.0%, they require special electric lines and water cooling. These flow-type gas lasers consume the gases SF_6, O_2, He, and H_2. Typically, the laser output consists of pulses with a duration of tens of nanoseconds, energy of tens of milli-joules per pulse, and repetition rates of several hertz. The average power output is typically less than 10 W. The CW HF lasers may emit tens of watts per line for the emission lines between 2.6 and 3.0 μm. The total power output may be more than 100 W.

2.3.3.3 New Semiconductor Lasers

Phased Arrays

The semiconductor lasers described earlier have a fundamental limitation: the power level emitted by one such laser is usually small (of the order of tens of milliwatts). This power is sufficient for optical communication but not for most medical applications.

The last few years have witnessed rapid progress in the development of arrays of lasers. Scientists learned to fabricate a large number (tens or hundreds) of semiconductor lasers on one substrate, so that the total power emitted by the array would be large. Studies found that a strong interaction between neighboring lasers results when the individual lasers are placed at a very small distance from each other. The radiation from one laser "leaks" into its two neighbor lasers. This interaction causes the individual lasers to start oscillating in a coherent fashion. In many respects, they behave as a large single laser. Such arrays of lasers are sometimes called phased arrays (like the phased array of radars used by the military). They have two characteristics that make them useful for medical applications:

(i) The power levels are relatively high, of the same order of magnitude as those of conventional lasers.
(ii) The beam behaves much like a conventional laser beam, allowing it to be focused into an optical fiber.

The total power available from an array of lasers is increasing; some can emit pulses of radiation with peak powers of more than 100 W or continuous radiation with average power of tens of watts.

Currently, the lasers emit light in the near-infrared region (0.8–1 μm). This light, which is not highly absorbed by tissue, may well be useful for "deep" heating such as hyperthermia cancer treatment. The wavelength of light emitted by the lasers can also be transmitted by optical fibers and used in laser catheter or laser endoscope applications, which will be discussed in later chapters.

Visible Semiconductor Lasers

There has been wide interest in shifting the laser light emission of semiconductor lasers from the near IR to the visible. It was found that lasers based on InGaP emit at 670 nm and those based on AlGaInP emit at 630 nm. After years of research and development, these lasers, with power output of several milliwatts, have become available commercially.

There is particular interest in the use of visible semiconductor lasers which emit high-power red light at around 630 nm. These lasers may eventually replace the metal vapor lasers for photodynamic cancer therapy (see Sections 3.7.2.1 and 9.7.3).

2.3.3.4 Free-Electron Laser

In the lasers that were described earlier, laser light is generated by excited electrons that are bound to atoms or molecules. There is another class of lasers in which light is generated by a stream of "free" electrons. This stream passes through an array of permanent magnets that are arranged periodically with their poles up or down. The interaction between the stream of electrons and the periodic magnetic field generates a beam of light along the path of the electron beam. A pair of mirrors added to the device generates a beam of laser light. This device is aptly called the free-electron laser (FEL).

This laser has several characteristics that are totally different from those of conventional lasers:

(i) The wavelength of the emitted light can, in principle, be varied continuously (i.e., tuned) over a very broad range. It is predicted that the wavelength of free-electron lasers will vary between the millimeter region and the extreme UV. Tuning is obtained by varying the energy of the stream of electrons or the spacings and the strength of the permanent magnets. Continuous tuning relies on the fact that we are dealing with free electrons, which do not have the same constraints as the electrons that are bound to atoms or molecules. Therefore, whereas bound electrons must jump from one quantum level to another, emitting a photon with specific energy (and therefore specific wavelength), free electrons can jump nearly any magnitude. The emitted photon may possess nearly any energy and therefore wavelength.

(ii) The efficiency of the free-electron lasers is potentially very high.

(iii) The free-electron lasers could potentially generate relatively high emitted laser power.

All these advantages make the free-electron laser a very appealing tool at the hand of the physician. Some physicians believe that, in the future, more than one laser will be needed for some procedures. While one laser was used for cutting, another could assist in the coagulation of blood. Instead of using different lasers, a single FEL laser could be used. A widely tunable laser will also be a powerful research tool. There are numerous cases in which physicians are eager to check the most efficient wavelength for a given treatment. At present, this research depends on the availability of a number of lasers. In the future, one FEL may be sufficient for research.

Present-day FELs are still very cumbersome to use as their size is similar to that of a small building. A team of scientists is needed to operate them. If smaller and less expensive FELs are developed in the future, they may become more accessible to the medical community. Yet their role will probably be limited to research, while smaller and cheaper lasers will actually be used for diagnosis and therapy.

2.3.3.5 Miscellaneous Laser Techniques

General techniques have been used for modifying the operation of several types of lasers. Such techniques may change the laser wavelength, add new lasing wavelengths, shorten the laser pulse length, and so forth. In this section we mention two of these techniques:

(i) Nonlinear effects: These effects occur when a high-intensity laser beam interacts with certain materials. The nonlinear interaction may give rise to the emission of other frequencies. Of particular interest are interactions in which an input laser beam of a given frequency is transmitted through a material and the output contains "new" frequencies which are exactly twice, three times, or four times the original frequency (the wavelength is shorter by a factor of 2, 3, or 4). These are called harmonics.

The crystal potassium dihydrogen phosphate (KDP) is a commonly used nonlinear crystal. Hydrogen may be replaced by deuterium or titanium (KTP) to improve their efficiency. For example when an Nd : YAG laser operating at a wavelength of 1064 nm is transmitted through such a crystal, one may obtain the wavelengths 532 nm (second harmonic), 355 nm (third harmonic), and even 266 nm (fourth harmonic).

(ii) Q switching: As explained in Section 2.2.1, laser operation is based on amplification provided by a gain medium and on feedback provided by mirrors. If either of these factors is sufficiently reduced by some mechanism, which could be termed "loss," lasing is terminated. It was shown that loss can be introduced into a laser for a period of time and when the loss is abruptly removed a short pulse (e.g., tens of nanoseconds) with a high energy per pulse (e.g., tens of millijoules) and a very high peak power (e.g., tens of megawatts) is generated. This method is called Q (for quality) switching and is widely used for generating short pulses.

2.3.4 Medical Laser Systems—Fundamentals

Lasers are the basic building block of a laser system. The physician or the surgeon needs a system that will be easy to operate and reliable. In this section, the basics of existing medical laser systems and their operation are briefly discussed (Hecht, 1992). Laser systems that make use of optical fibers are discussed in Chapter 8.

Most laser systems can be divided into three subsystems: the laser itself, a beam delivery unit, and auxiliary subsystems.

2.3.4.1 The Laser

The laser itself consists of two parts: the head and the power supply.

(i) *Laser head*: This section can be divided in two—the lasing medium (and its container) and the two mirrors. The lasing (i.e., active) medium is a gas, liquid, or solid. With gases or liquids, the laser medium must be stored in a container. This is a weak spot in the system. Although the manufacturer provides so-called leakproof joints, the insubordinate gases and liquids have a natural tendency to leak from seals and joints. This is particularly troublesome with highly toxic gases such as those present in excimer lasers. Gases and liquids are often circulated through the system. Pumps are therefore included in such a system, and these can be noisy. Over the past decade, there has been constant progress in the reliability of gas and liquid circulating subsystems.

The two mirrors that form the feedback system of the laser are sometimes placed on special mounts, separate from the laser medium. These mirrors may also be attached to the laser cavity. In lasers such as the HeNe laser, in which the mirrors are glued or welded to the cavity, the mirrors cannot move and lose alignment.

Not all of the electrical energy required for laser operation is converted into light; a large fraction is wasted as heat. In order to get rid of this heat, the laser head often has to be cooled by water circulation. Unimpeded flow of cooling water is essential for effective laser operation. Naturally, such a system requires regular maintenance.

(ii) *Power supply*: Most lasers need a special power supply. For example, gas lasers require a high voltage to start a discharge (much as in a fluorescent lamp at home). The same power supply then provides a constant current to operate the laser. In many cases, a rather large current (and a large amount of energy) is needed to operate the laser. For this reason, a three-phase supply is often necessary. In many laser systems the power supply is housed in a separate console together with pumps, gas bottles, and all the laser controls. This is convenient for the laser manufacturer but not necessarily for the laser user.

2.3.4.2 Beam Delivery Unit

The laser beam may be guided to the operating site by an articulated arm, as discussed below, or by optical fibers, discussed in Chapters 4 and 8.

The laser beam is a unidirectional beam consisting of an ordered series of waves—all with the same wavelength and moving in one direction. In order to steer the beam from one place to another, mirrors are often used. With surgical lasers, the beam may be directed toward the operating site using the articulated arm, which consists of several (six to eight) mirrors that are connected to each other by a set of tubes, as shown in Fig. 2.17. Each mirror is mounted on a rotating holder. By moving the tubes, the angles between the mirrors are changed. A laser beam is directed toward the first mirror and then reflected from mirror to mirror. By moving the whole arm, one may change the position and the direction of the beam that emerges from the articulated arm. The physician thus has the flexibility to deliver the laser beam from the laser head to the operating site. At the end (handpiece) of the articulating arm, there is usually a focusing lens that is used to focus the laser beam onto the desired spot. By moving the handpiece, the physician can move the focused laser beam from one spot to another.

Unlike the laser, it is not easy to realign an articulating arm. Such alignment is normally done by the manufacturer. In the past, articulated arms were fairly heavy and difficult to control. They were also easily moved out of alignment. In recent years, there has been progress in fabricating lightweight arms that are robust and reliable. Many articulated arms will probably be replaced in the future

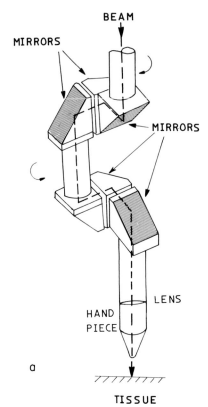

FIGURE 2.17   (a) Laser beam manipulation by mirrors—the articulated arm. (b) Medical CO_2 laser with an articulated arm (*facing page*). (Courtesy of Laser Industries.)

by special optical fibers that can transmit high power levels (more than 100 W continuously). For some applications, such as the delivery of CO_2 laser beams or for other wavelengths that cannot be easily transmitted through optical fibers, articulating arms will still be required.

2.3.4.3 The Auxiliary Subsystems

The beam that exits the articulated arm may be further manipulated by another optical subsystem. For example, the distal tip of the articulated arm may be connected to an operating microscope, so that the physician can perform laser microsurgery. This is shown in Fig. 2.18. Alternatively, the distal tip may be connected to a rigid endoscope. The laser beam is focused through the endoscope with a long-focal-length lens. This makes it possible to carry out endoscopic laser surgery. This laser endoscope is shown in Fig. 2.19.

The first operable lasers were developed by scientists for research and fun. In a scientific laboratory, the basic requirements for a laser system are different from

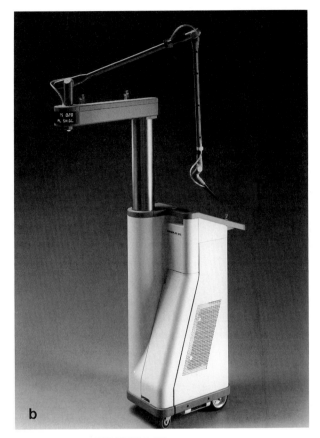

b

FIGURE 2.17 *(Continued)*

the requirements of a surgeon in the operating room. When lasers were first used in medicine, manufacturers used the "scientific" lasers with very few modifications (except in price!). With the increased use of lasers for medical applications, manufacturers have made efforts to adapt the laser to fit the hospital and the physician's office environment. Although laser systems available today are still not as reliable as some physicians would like them to be, there have been tremendous improvements. The systems are very likely to continue to improve.

2.3.5 Medical Laser Systems—Principles

The design problems involved in the production of the various medical lasers have not been easy to overcome. Lasers produce a pencil-thin beam of light, and in order to use this beam for diagnosis or therapy, it must be directed toward the pertinent region. Delivering laser light to tissue is at least as difficult as producing the light in the first place. Although it may seem relevant only to the optical en-

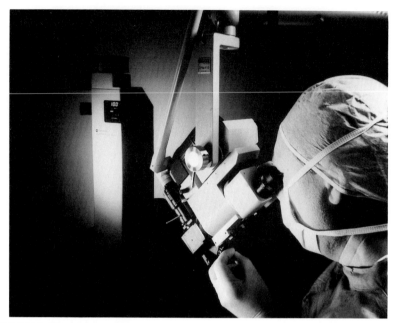

FIGURE 2.18 CO_2 laser connected to an operating microscope via an articulated arm. (Courtesy of Coherent Inc.)

gineer at the factory, physicians, technicians, and nurses who operate the laser equipment should also understand the basic principles underlying the delivery and control of laser beams. A few points on the laser beam itself are appropriate at this stage.

Gaussian Beam

A possible misconception must be dealt with at this stage in order to alleviate any later confusion. Although the laser beam appears uniform to the naked eye, the intensity is, in fact, strongest at the center of the beam and decreases gradually away from the beam center. The distribution of the beam intensity is often Gaussian or bell shaped, as discussed in Section 2.2.3.2. This point is emphasized because the cutting action of the laser beam depends on high intensity. Cutting is thus most efficient in the center of the beam; toward the edges the beam may not cut at all.

Focusing

The beam emitted by the laser usually has a diameter of several millimeters. If this beam impinges on tissue, it will be absorbed. Although the tissue will be heated, the power density may not be high enough to vaporize tissue. In order to cut tissue, the beam must first be focused by means of a lens to a much smaller spot (see Section 2.2.3). The major difference between laser light and regular light is that, in principle, a Gaussian laser beam can be focused to a spot whose diameter is equal to the wavelength of the laser. If, for example, a laser beam whose wave-

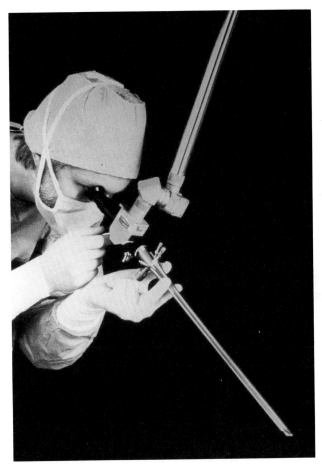

FIGURE 2.19 CO$_2$ laser connected to a rigid endoscope via an articulated arm. (Courtesy of Laser Industries.)

length is 10 μm (e.g., CO$_2$) is under consideration, the whole laser beam can be theoretically focused onto a spot whose diameter is also 10 μm. In practice, the lenses are not ideal and the actual spot size may have a radius of tens of micrometers. The power density in this spot will still be extremely high and the laser beam will be able to vaporize tissue at the focal spot.

Misalignment

The laser beam does not always obey the theoretical prediction. As already mentioned, the laser consists of a lasing medium and two mirrors. The shape of the output beam is highly dependent on the alignment of the two mirrors. If they are even slightly out of alignment, the intensity distribution is modified. The distribution may look like two bells or four bells side by side, as shown in Fig. 2.8, or an irregular shape as shown in Fig. 2.20.

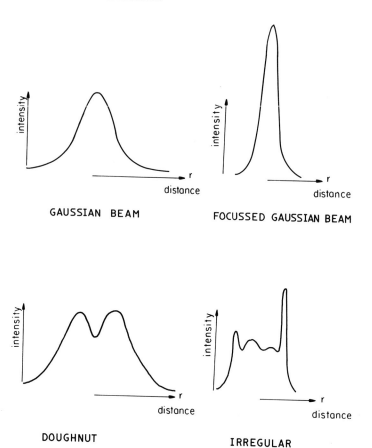

FIGURE 2.20   Regular (Gaussian) and irregular beam shapes.

Why should this peculiar behavior be emphasized? Simply, because the non-Gaussian beam will not focus to a small spot. The focused beam will spread, dramatically reducing the power intensity at the focal spot and the cutting efficiency of the beam. Earlier lasers could easily be knocked out of alignment by moving the laser system from one room to another. Because mirror mounts are now constructed more ruggedly, such mishaps are currently rare. Misalignment is, however, not a major disaster and can be corrected by a qualified person. Lasers should still be handled with care.

2.4 LASER SAFETY—FUNDAMENTALS

The issue of safety is extremely important when working with lasers. Many aspects of laser safety are discussed in the detailed book by Sliney and Wolbarsht (1980).

2.4.1 Optical

The unique properties of laser light are suitable not only for diagnosis and therapy but also for unwanted damage to human tissue. Of all the body tissues, the retina in the eye is the most vulnerable to laser light. Accidental overexposure may cause severe damage. Special attention should be paid to protecting the eyes against such exposure. The structure of the eye is shown schematically in Fig. 2.21.

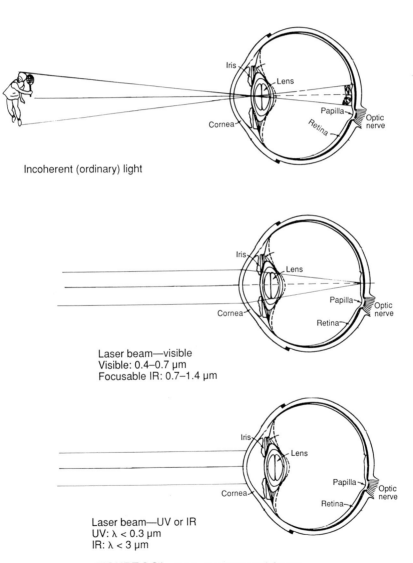

Incoherent (ordinary) light

Laser beam—visible
Visible: 0.4–0.7 μm
Focusable IR: 0.7–1.4 μm

Laser beam—UV or IR
UV: λ < 0.3 μm
IR: λ < 3 μm

FIGURE 2.21 Safety: laser beam and the eye.

In observing an object (using ordinary, *incoherent* light), the eye forms a reduced image of the object on the retina, as shown in Fig. 2.21a. This is done with the help of the cornea and the lens, both of which are highly transparent to visible light. This imaging process is harmless to the eye. If visible *laser* light reaches the eye, it is focused by the lens into an extremely small spot (see Fig. 2.21b) and the power density in this spot is often very high.

EXAMPLE: Let us consider an HeNe laser beam that impinges on the eye and is focused on the retina. The focusing system of the eye has a focal length of about $f = 20$ mm. Therefore the spot size on the retina, calculated from Eq. (2.7) has a diameter of about 12 μm. If the laser beam power is 5 mW and the radius of the beam near the eye $w = 1$ mm $= 0.1$ cm (diameter of 2 mm), the power density of the beam is $[0.005 \text{ W}/\pi(0.1)^2] = 1.7 \text{ W/cm}^2$. This is actually the power density on the cornea of the eye. On the retina, however, the power is roughly the same, but the spot diameter is 12 μm (and radius 6 μm). The power density is now $[0.005/\pi(0.0006)^2] = 5000 \text{ W/cm}^2$. The power density on the retina is increased by a factor of 3000. Roughly 5% of that power is absorbed in the rods and the cones in the retina. Since the retina is built to respond to low power density levels, it may be damaged by such high-level intensities. This may lead to loss of vision.

The visual system of the eye does not respond to wavelengths in the spectral region $0.7 - 1.4$ μm in the near IR. Yet the elements of the optical system of the eye (cornea, lens, and the eye liquids) are transparent at these wavelengths. The near IR is therefore, in a sense, more dangerous because we cannot see the laser light, which nevertheless may be focused on the retina and cause damage.

Laser wavelengths in the far UV (<0.3 μm) or the mid-IR (>3 μm) are highly absorbed in tissue. A laser beam at these wavelengths will therefore be totally absorbed in the cornea and will not reach the retina, as shown in Fig. 2.21c. Controlled absorption of this nature is the basis for laser keratotomy, a novel ophthalmological procedure for ablating part of the cornea and thereby correcting nearsightedness. On the other hand, accidental exposure to these wavelengths may cause damage to the cornea.

The laser beam used for surgical applications or for diagnosis is a narrow and intense beam. It may easily be reflected by surfaces that are characterized by smoothness, such as shiny metal or polished glass. Such a reflection is called a specular reflection. As a result, a laser beam may inadvertently be reflected by some surfaces, in particular from metallic instruments, into the eye. The amount of energy reaching the eye may be sufficient to cause damage. This is true for visible, UV, and IR lasers. When a laser beam strikes a rough surface, the reflected beam is scattered in all directions, resulting in a diffusely reflected beam. The beam reflected from human tissue, from dull metal and nonsmooth surfaces, or from cotton is diffusely reflected. Its low intensity makes it much less dangerous to the eye.

Extensive work has been carried out to determine the maximum laser energy that can impinge on the eye without causing damage. Experiments were done for

TABLE 2.4 **Safety Classification of Lasers**

Class I	Powers do not exceed the maximum permissible exposure (MPE) for the eye.
Class II	Visible laser beams with power levels up to 1 mW. The eye is protected by the blinking reflex.
Class IIIa	Same as class II but used for power levels up to 5 mW and with laser beams which are expanded.
Class IIIb	Power level up to 0.5 W. Direct viewing of the beam is dangerous.
Class IV	Power levels >0.5 W. Extremely hazardous.

the various wavelengths, under pulsed and CW conditions and for various power densities (Sliney and Wolbarsht, 1980). In each case, a maximum permissible exposure (MPE) was obtained and the data compiled in MPE tables. Several government agencies used these tables to classify lasers according to the danger they present to the eye. This internationally accepted classification is shown in Table 2.4.

EXAMPLES: The HeNe aiming laser beams in IR systems (e.g., CO_2 or Nd: YAG) with power output up to 1 mW are in class I. Those with power output up to 5 mW are in class II. Visible gas lasers, which are used for diagnosis, or semiconductor lasers with power output less than 0.5 W are in class IIIb. Most of the lasers used for laser surgery and therapy have power output more than 0.5 W and are therefore in class IV.

Dangers to the eyes can be avoided with care. Physicians, personnel, and patients who may be exposed to laser beams must wear appropriate protective goggles. Special goggles are made of plastic materials which serve as filters that "block" the particular wavelength (color) emitted by the laser. They transmit other colors—so it is possible to see through the goggles—but they do not permit the transmission of laser light into the eye. The goggles are rugged and are securely fitted around the eye or over prescription glasses to protect against stray laser beams entering from the side. The blocking filter may also be attached to the eyepiece of optical systems, such as binocular microscopes or endoscopes, that are used in laser surgery and therapy.

The other organ vulnerable to laser injury is the skin. The beam from intense CO_2 or Nd: YAG lasers may cause burns. In the case of excimer lasers, there are indications that the radiation at 248 nm may be carcinogenic. Each case requires complete control of the beam.

2.4.2 Smoke

Another point to consider is the "smoke" generated when a laser beam cuts tissue. The smoke is unpleasant and may also be carcinogenic. Furthermore, some reports have indicated that viable viruses can exist in the smoke. Smoke can also absorb and block some of the laser output from reaching the tissue. It can be evacuated from the operating site using extractor fans. This elementary procedure must be observed when performing laser surgery.

2.4.3 Electrical

Lasers are large electrical instrument systems with inherent electrical danger. The only fatal accidents resulted from electrocution, except for one reported fatality caused by the laser falling on top of its operator. Most gas lasers have power supplies that run on 220 V to provide high voltage for operating the laser. Because these power supplies are very well insulated, they present almost no danger to the operator. If anything goes wrong, they must be handled only by qualified personnel. Incidentally, this is also true for many domestic systems such as TV sets or microwave ovens.

References

Hecht, J. (1992). *The Laser Guidebook*, 2nd ed. New York: McGraw-Hill.

Meyers, R. A. (1991). *Encyclopedia of Lasers and Optical Technology*. New York: Academic Press.

Siegman, A. E. (1986). *Lasers*. Mill Valley, CA: University Science Books.

Sliney, D., and Wolbarsht, M. (1980). *Safety with Lasers and Other Optical Sources*. New York: Plenum Publishing.

Wilson, J., and Hawkes, J. F. B. (1987). *Lasers: Principles and Applications*. Englewood Cliffs, NJ: Prentice Hall.

Winburn, D. C. (1987). *Lasers*. New York: Marcel Dekker.

Yariv, A. (1991). *Optical Electronics*, 4th ed. Philadelphia: Holt, Reinhart & Winston.

3

Applications of Lasers in Therapy and Diagnosis

3.1 INTRODUCTION

Photobiology is the study of the interaction between electromagnetic radiation (e.g., the optical spectral range 400–700 nm) and biological molecules. It thus involves the study of photochemical reactions caused by light and the resulting biological responses. Photomedicine utilizes the results of these studies for medical diagnosis and therapy. These applications have been discussed in several books (Atsumi, 1983; Martellucci and Chester, 1989; Pratesi, 1991; Regan and Parrish, 1982) and review articles (Berns, 1991; Parrish and Deutch, 1984). In addition, special issues of scientific journals are dedicated to this topic (Alfano and Doukas, 1984; Birngruber, 1990; Caro and Choy, 1992; Deutch and Puliafito, 1987; Deutch and Wynant, 1992; Welch, 1989; Wilson, 1991).

Light may be absorbed in the skin and converted to heat. The *thermal* effects of this process may be used for therapy and have been in use for millennia. Greek physicians recommended sunbathing for the cure of various diseases. More interesting, however, is the use of *nonthermal* effects of light. Almost 4000 years ago, ancient Egyptians used nonthermal phototherapy (Edelson, 1988). They noticed that people who ate certain plants (i.e., *Ammi majus* or *Psoralea corylifolia*) that grew on the banks of the Nile became more susceptible to sunburn. They used this information to treat vitiligo, a skin disease in which areas of the skin lose their pigmentation. The active ingredients in those plants, psoralens, are used in modern photomedicine. Today they are triggered by laser beams—the topic of this chapter.

Useful artificial light sources were developed in the late 18th and early 19th centuries, at the same time that invisible rays of light (i.e., infrared and ultraviolet) were being discovered. The heating effects of incandescent light sources were used for therapeutic applications. Tyndall (who is also mentioned in Chapter 4) found that visible rays of light generate heat deeper in the body than invisible ultraviolet rays. This was confirmed later (see Section 3.4), in the 19th century, by Finsen, who found that invisible ultraviolet rays from the sun or from artificial sources also have therapeutic effects.

Soon after the development of the first lasers in the early 1960s, physicians began to investigate the possible therapeutic uses of these new light sources. One of the organs that lends itself readily to investigation is the skin. Leon Goldman, a dermatologist from Cincinnati, Ohio, led pioneering studies in laser surgery and therapy that served as a beacon for further study. Shortly thereafter, laser light was focused into the eye and used to produce burn scars that "weld" detached retina to the underlying choroid. Local heating inside the eye was also used for the treatment of vascular disorders such as proliferation of blood vessels. One of the more interesting works of that era is the short paper by McGuff *et al.* (1963) that discusses the use of lasers for the treatment of cancer and cardiovascular disease. Another interesting paper, by Yahr and Strully (1966), discusses the various applications of lasers in medicine. They mention the possible use of optical fibers both for imaging and for transmitting laser energy for internal operations, in particular for brain surgery and for the fracture of stones that block biological duct systems. The specific applications mentioned in both papers are today among the most promising uses of lasers in medicine.

During the last three decades, lasers have gradually penetrated medical practice. From a slow start during the first two decades, faster growth has occurred in the last decade (Carruth and McKenzie, 1986; Council on Scientific Affairs, 1986; Dixon, 1987; Goldman, 1991; Parrish and Wilson, 1991; Oster, 1986). This chapter discusses the basic interactions between lasers and tissue and their applications for both diagnosis and therapy. The more clinical aspects of laser medicine are given in Chapter 9.

It is the opinion of this author that medical laser systems will be integrated with optical fibers and endoscopes (see Chapter 8). This chapter discusses mostly the aspects of laser-assisted diagnosis and therapy that are closely linked to optical fibers (Bown, 1986).

3.2 LASER-ASSISTED DIAGNOSIS AND THERAPY—FUNDAMENTALS

Chapter 2 examined some of the unique properties of laser light. The use of lasers for therapy depends on the interaction of laser beams with tissue. Several types of interactions may take place, depending on the wavelength of the laser, temporal nature of the beam (continuous wave or pulsed), energy delivered, and spatial nature of the beam (focused or unfocused).

3.2.1 Laser Interaction with Tissue

3.2.1.1 Reflection, Absorption, and Scattering

When light is incident on a sample of material, such as a colored window, it is partially transmitted and partially attenuated through the material. Attenuation is caused by three distinct processes: reflection, absorption, and scattering. Reflection occurs at the surface of the sample. When the surface is very smooth, we observe "mirrorlike" reflection from this surface. Absorption in colored windows is often due to coloring agents inside the sample. The absorbed light is generally converted to heat. Scattering causes the light to spread in different directions. Scattering may occur inside the material and is obvious, for example, when light travels through fog or smoke. It may also occur on the surface of a sample, as in the case of a "frosted" window. All these effects are highly dependent on the wavelength of the incident light. For example, an ordinary home window is highly transparent to visible light but almost completely absorbs ultraviolet (UV) light at wavelengths $\lambda < 300$ nm or infrared (IR) light at $\lambda > 3$ μm.

Light incident on biological tissue is attenuated because of the same effects. There is reflection from the surface of tissue (e.g., the skin) and strong scattering occurs inside the tissue. Absorption results from the chromophores (coloring agents), such as water, hemoglobin, and melanin. Absorption is also highly dependent on wavelength. Whereas water absorbs in the UV and IR parts of the optical spectrum, it transmits well in the visible. Hemoglobin absorbs in both the UV and the blue–green part of the spectrum ($\lambda < 0.6$ μm) but does not absorb in the red part. This can easily be demonstrated if light from a high-intensity lamp is transmitted through a hand. The hand looks red because the green blue light is absorbed and only the red light is transmitted. The blue light emitted by Ar ion lasers is transmitted by water in tissue but is highly absorbed by hemoglobin. Additional chromophores, such as the melanin pigment in the skin, absorb throughout the visible and the near-infrared parts of the spectrum. On the other hand, the IR light emitted by the CO_2 laser is highly absorbed by water (and therefore most types of tissue).

This section discusses two types of interactions between light and tissue. These interactions are thermal and nonthermal, and each may be used for diagnosis or therapy.

3.2.1.2 Thermal Interaction

The thermal interactions between laser beams and materials have been carefully studied by scientists and engineers over the last two decades. The heating of materials by lasers is widely used in industry for heat treatment of metals and for welding and cutting. Although the basic theory behind the interaction of lasers with tissue is quite similar, the results are somewhat different. Primarily, we normally deal with much lower temperatures than those used in material processing. Second, effects which are particular to living tissue, such as coagulation, thermal damage, and tissue vaporization, are considered. In this section, only the fundamentals of thermal interaction with tissue are described.

When a laser beam impinges on tissue, some of the energy is scattered and some is absorbed. The energy absorbed in the tissue generates heat and causes a temperature rise. If the laser beam is turned off, the temperature decreases again. The theoretical work done in material science enables one to calculate the variation of temperature with time, $T(t)$, during the heating and cooling processes.

If the temperature rise is modest, that is, the final temperature is below 40°C, only nondamaging effects are normally observed. When the temperature increases to higher values and is kept there for an extended period, thermal damage to cells may occur. If the temperature rises above 100°C, water starts to evaporate. As long as there is water in tissue the temperature cannot normally rise above 100°C (as in the case of boiling water at atmospheric pressure). Once all the water has evaporated, the temperature will rise to a higher temperature T_A, called the ablation temperature, where other components of the tissue evaporate. This is the process that leads to tissue removal.

In laser surgery, a laser beam is focused onto a small spot. If the power density of the focused beam is high, tissue at the focal spot will rise to temperatures higher than T_A. Tissue will be removed and a crater will be formed. Simultaneously, the temperature of neighboring tissue will also rise. In areas close to the crater, the temperature often rises to values that cause both thermal damage to tissue and blood coagulation. In more distant areas, the temperature rise is limited and no thermal damage is observed. This thermal behavior is normally observed in histological samples obtained from tissue that is cut by a laser beam (see Section 3.7). Another type of interaction is the generation of plasma (i.e., ionized gas at an extremely high temperature) and shock waves. This happens when an area of tissue is exposed to a very short laser pulse of extremely high power density. The shock wave may assist in removing hard tissue.

3.2.1.3 Nonthermal Interaction

There are several cases in which the interaction between laser beams and tissue is not thermal. Photodynamic therapy is based on the introduction into tissue of chemical substances that can absorb light of a specific wavelength. These sensitizers are normally inert. If they absorb this specific wavelength, however, a chemical change occurs and the substance interacts chemically with the surrounding tissue. Laser light can therefore be used for triggering drugs inside the body.

Another example is the interaction of excimer lasers with tissue. Histology shows that the beam removes tissue with practically no thermal damage. It is postulated, and good experimental evidence exists to show, that the excimer ablation occurs without vaporization and that some UV laser ablation occurs by mechanical ejection of nonvaporized tissue.

3.2.2 Laser-Assisted Diagnosis

Lasers have been used in a variety of diagnostic techniques in physics and chemistry. The same techniques make lasers useful for diagnostics in biomedical applications. Diagnostic methods that are compatible with fiberoptics are discussed below.

One simple method is based on the reflection of light from tissue. Diseased tissue (e.g., tumor) often looks different from healthy tissue. This could be determined quantitatively by illuminating both tissues with light of a given power and wavelength in order to measure the amount of reflected light at various wavelengths. The differences between reflection spectra are used to distinguish the two types of tissue. Another common diagnostic technique is based on luminescence. Light illuminates tissue and luminescent light of a characteristic wavelength is emitted. This is illustrated by two examples:

Luminescence in the dark: At night, a car's headlamps illuminate the road. Regular objects simply reflect the headlamp light. Luminescent road signs absorb the white light and emit a different, characteristic light (e.g., red light in stop signs) which make them more visible. UV light sources, known as black light, are used at parties or nightclubs to demonstrate a similar effect. The UV light, which is invisible to the eye, is partially reflected from most objects, but it cannot be seen. The UV light is also absorbed in several materials that have a strong visible luminescence, such as certain plastic threads. This luminescence is obvious in the dark. The ordinary UV light therefore serves as a diagnostic tool for detecting plastic fabrics in the dark. Similar methods are used for tissue diagnosis.

Wood's lamp: This is based on a high-pressure mercury lamp whose light is transmitted through a glass filter. It transmits light only in a narrow spectral region about 365 nm. Wood's lamp is used in dermatology to detect fungal or bacterial infection. The infected tissue often fluoresces with a yellow or orange color.

The laser has a similar role that is especially important in cases in which simple observation does not reveal differences between tissues. Under white light for example, malignant tissue may look identical to healthy tissue. If excited by UV laser light, however, the tissues may have different luminescence. If so, this may be a powerful tool for distinguishing between malignant and normal tissue. Moreover, this diagnostic procedure lends itself to easy use in endoscopy. One may excite tissue by UV light through an endoscope and look for the characteristic emission of malignant tumors. This and other fiberoptic diagnostic methods are discussed in more detail in the following chapters.

3.2.3 Therapy by Thermal Interaction between Lasers and Tissue

Most of the tissues in our body are not transparent to light; that is, light is absorbed in such tissues. The absorption process converts the energy of the light into heat. Different lasers heat tissue in different ways, depending on the absorption characteristics of the various wavelengths.

(i) *Different wavelengths*: For some wavelengths (i.e., UV or middle and far IR), the laser beam is highly absorbed in tissue. All the energy is therefore absorbed in a thin layer near the surface, where rapid heating occurs. For other wavelengths the absorption is small (i.e., Nd : YAG laser), resulting in slower heating of a larger volume of tissue.

(ii) *Power density*: It is expected that the temperature rise in the tissue will depend on the power density (the ratio between the beam power and the irradiated area). For low power densities, minimal heating is observed. At higher power densities, vaporization of water and tissue removal are expected.

(iii) *Duration*: The same amount of energy may be delivered to tissue continuously, in very long pulses or in very short pulses of high peak power. The effects on tissue are different, as discussed below.

Indeed, heat generated by an absorbed laser beam has been widely used for therapy. Some of the important applications are discussed here:

(i) *Biostimulation*: There are claims that the minute heat generated by low-power laser light stimulates nerves, accelerates healing of wounds, and reduces pain.

(ii) *Hyperthermia*: This is a therapeutic modality for cancer that depends on heating of malignant tissue to $42-45°C$. Heating may be performed by a laser beam that penetrates deep into tissue (e.g., Nd : YAG).

(iii) *Thermotherapy*: Heating tissues to temperatures above $45°C$ causes tissue necrosis and destruction, as in the laser treatment of cancer or enlarged prostate glands.

(iv) *Coagulation*: If the laser beam is absorbed in blood, it may coagulate the blood. Coagulation is vital in sealing blood vessels and stopping bleeding during surgery.

(v) *Welding*: The absorbed laser heat facilitates the joining of tissues and, in particular, connects blood vessels to each other (i.e., anastomosis).

(vi) *Vaporization*: When the power density is sufficient, the temperature of the tissue rises above the ablation temperature, causing the tissue to evaporate. Vaporization is the basis for laser surgery. It is thus advantageous to use a wavelength that is highly absorbed in tissue (e.g., CO_2 or excimer laser)

(vii) *Shock waves*: Short pulses of high peak power that are absorbed in tissue may generate plasma and often generate shock waves. These shock waves have been used in ophthalmology to remove tissue inside the eye and to fragment urinary and biliary stones.

3.2.4 Therapy by Nonthermal Interaction of Laser Beams and Tissue

Some techniques rely on a nonthermal interaction between laser light and the tissue. At relatively low energies, phototherapy may be performed by triggering chemical reactions in tissue. At relatively high energies, direct ablation of tissue by photons (photoablation) may occur.

(i) *Photochemical interaction*: The most widely studied case of this nature is cancer treatment by photodynamic therapy (PDT, also called photoradiation therapy). The compound hematoporphyrin derivative (HPD) is one of many photosensitizing agents which may be incorporated in malignant cells. When exposed to specific wavelengths, this compound stimulates the production of singlet oxygen, which kills the cancer cells. A high light irradiance is needed in order to obtain efficient therapy. Both requirements are not met by an ordinary light source, but rather by a laser.

(ii) *Photoablation*: When tissue is exposed to the focused beam of an excimer laser, one observes a clean cut. Tissue is removed without any noticeable thermal damage to the side walls of the cut. This is different from damage caused by other lasers. It is claimed that the lack of thermal damage depends on the fact that excimer lasers operate in the UV part of the spectrum. It is suggested that the energetic photons give rise to tissue removal through a direct process that does not involve heat. This conclusion has not yet been verified, since other lasers such as CO_2 or Er:YAG lasers cause similar effects, although the photon energies at these IR wavelengths are low.

The clean cut generated by the excimer laser is important in eye surgery. The excimer laser is used for an operation called photorefractive keratectomy, which reshapes the cornea and corrects its refractive properties. The major advantage of the excimer laser is that the transparency of the cornea is not impaired during the procedure. This procedure is still in experimental stages.

3.3 INTERACTION OF LASER BEAMS AND MATERIALS—PRINCIPLES

In this section we discuss the optical aspects of the interactions between laser beams and materials in general. The interactions between laser beams and biological tissue are discussed in Section 3.4.

3.3.1 Transmission of Laser Beams through Materials

When a collimated light beam impinges on a thick sample, only a fraction of the light energy is transmitted through the sample. The rest is lost by three processes: reflection, absorption, and scattering (Hecht, 1987). This is illustrated schematically in Fig. 3.1, where a collimated beam of irradiance[5] (or intensity) I_i is shown incident on a planar slab of material. The various loss processes that can occur are represented by the arrows I_r (reflected irradiance), I_s (scattered irradiance), I_a (absorbed irradiance), and finally I_t (transmitted irradiance). The lengths of the arrows are meant to represent their magnitude. I_r includes the reflection

---

[5]Irradiance is the correct term in physics. Sometimes the term intensity is loosely used instead, although strictly speaking "intensity" refers to another quantity (see Chapter 10).

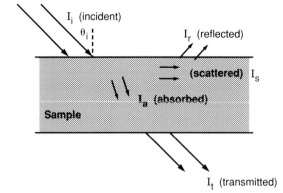

FIGURE 3.1 Reflection, absorption, scattering, and transmission in a sample.

from the front surface, while that from the back surface is neglected. Energy conservation requires that

$$I_i = I_r + I_s + I_a + I_t. \tag{3.1}$$

The three attenuation processes are discussed below. One should note that, for simplification, the discussion concerns the attenuation of a collimated beam of light. This applies to a laser beam (which is inherently collimated) or to a collimated beam of ordinary light.

Reflection

Optical materials can be characterized by an index of refraction. This is an inverse measure of the velocity of light in these materials, with respect to the velocity of light in vacuum. The index of refraction of air is approximately 1; the velocity of light in air is equal to that in a vacuum. The index of refraction of water is 1.33 and that of glass about 1.5, so that the light velocity in these materials is lower than that in air.

Suppose that a low-irradiance visible laser beam propagates in air and impinges on inorganic material. Assume that the material has an index of refraction n, that the surface of the material is planar and optically polished, and that the beam is not absorbed by the material. If the irradiance of the incident beam is I_i, there will be a reflected beam of irradiance I_r back into air and a refracted beam of irradiance I_t into the material. The ratio I_r/I_i is called *reflectance*, and for a beam that is perpendicular to the surface (normal incidence) it can be shown that at the interface $I_r/I_i = [(n - 1)/(n + 1)]^2$. This formula also holds for light that is incident from the material into air.

EXAMPLE: Light is incident from air on a slab of glass (e.g., a glass window) whose index of refraction is 1.5. In this case light is incident first from air onto glass; part is reflected and part is transmitted. We thus write $I_r/I_i = [(1.5 - 1)/(1.5 + 1)]^2 = 0.04$, so that 4% of the light is reflected back to air. Let us assume that the window is thin and that there is no attenuation due to absorption or scattering. The light traveling inside the window is incident on the glass–air interface; part is reflected back into the glass and part is transmitted to air. There is again a 4% reflection loss in the glass–air interface. The total reflection loss is thus 4% + 4% = 8% and 92% of the light is transmitted through the window.

Absorption

The beam that is propagating through a material is very often absorbed by the material. The light may interact with the material in a variety of ways, including excitation of electronic transitions and molecular vibrations. In most cases, the attenuation of a laser beam, like all light beams, increases exponentially with the length l of the beam path in the material. It can be written in the well known form (Beer–Lambert equation)

$$I(x) = I(0) \exp(-\alpha x), \tag{3.2}$$

where $I(0) = I_i$ is the irradiance at $x = 0$ (the surface of the absorbing material) and $I(x)$ is the irradiance after a path length of x units. The factor α is called *absorption coefficient*. The behavior of $I(x)$ is shown in Figs. 3.2 and 3.3.

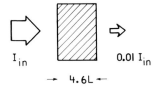

FIGURE 3.2   Light decrease in a sample due to absorption.

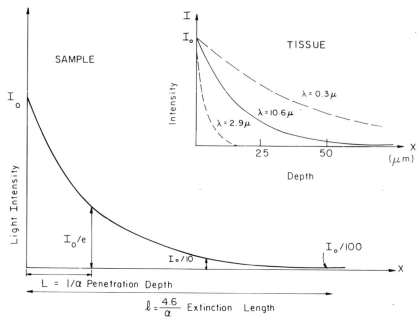

FIGURE 3.3 Penetration depth and extinction coefficient in a sample.

The units of the absorption coefficient are cm^{-1}. Thus αx is a dimensionless quantity. Assume that $I(0)$ is the irradiance of the incident (input) beam and that the absorption coefficient is $\alpha = 2$ cm^{-1}. After the beam passes through a slab of material with a thickness of $L = 1/\alpha = 1/2$ cm, the irradiance of the output beam will be $I(1) = I(0) \exp(-1) \approx I(0)/3$. The length $l = 1/\alpha$, at which the irradiance is decreased by a factor of about 3, is called the *penetration depth*. Similarly, for a layer with a thickness of $l = 2.2$ cm, the irradiance of the transmitted (output) beam is $I(2.2) = I(0) \exp(-2 \times 2.2) = I(0) \times 0.01$. In this case, 99% of the light is absorbed in the material and only 1% is transmitted. The thickness of the material required to absorb 99% of the incident radiation is often called the *extinction length* and is designated as l. In this example, $l = 2.2$ cm. The absorption coefficient and the extinction length are simply related, $l = 4.6/\alpha = 4.6L$ as explained in Figs. 3.2 and 3.3.

For a given material, the extinction length is often highly dependent on the wavelength of the incident beam. Regular silica-based glass, for example, transmits well in the visible part of the spectrum. If the incident beam is in the visible part of the spectrum, the extinction length of such glass for visible light may be tens or hundreds of centimeters. Regular glass, however, does not transmit well in the UV (see Fig. 4.13). If the wavelength is $\lambda < 0.3$ μm, the extinction length may be less than 1 mm. The same is true in the infrared, for wavelengths of $\lambda > 3$ μm. Ninety-nine percent of incident visible light will be absorbed only after passing through several meters of glass. A glass layer less than 1 mm thick will, however, absorb 99% of UV light or mid-IR light. The absorption of laser beams of different wavelength in tissue is important for laser surgery and it is illustrated schematically in the insert in Fig. 3.3. The extinction length is clearly dependent on the wavelength λ.

Another example is the sunscreen preparations which serve as UV-absorbing filters. The sun-blocking chemicals are creams, lotions, or gels which contain light-absorbing compounds, such as para aminobenzoic acid (PABA). Although these highly absorb the harmful UVA (320–400 nm) and the UVB (290–320 nm) solar radiation, they transmit visible light. Sunscreens therefore prevent the deleterious effects of excessive exposure of the skin to sunlight, while still being cosmetically acceptable.

Scattering

The absorption of light involves the conversion of the light energy into some other form of energy such as heat. However, not all the light is absorbed. Some of the light passing through a material is scattered inside the material or on its surface. Scattering means that a fraction of the beam does not continue in its original straight path but is deflected to other directions. The scattered light undergoes further reflections and scattering, until it is either absorbed (and converted to heat) or escapes from the sample. We may distinguish between two phenomena:

Surface scattering: Let us consider a glass sample with a surface that is rough and not well polished. A rough surface may be defined as one with troughs and valleys which are slightly larger in size than the wavelength of light. For visible light, the roughness is about 1 μm. In this case, light scattering can be observed. In addition to mirrorlike reflections, scattering in other directions is obtained. This decreases the amount of light that is transmitted into the material (or through the window).

Bulk scattering: This may result from some defects in the material, such as bubbles, cracks, or impurities. Some scattering mechanisms also exist in pure homogeneous material, due to the motion of the constituents.

The attenuation of a collimated beam due to scattering may often be approximated by the formula $I(x) = I(0) \exp(-\beta x)$. In this case, the coefficient β is called the *scattering coefficient*. If both absorption and scattering cause attenuation, we may denote $\gamma = \alpha + \beta$ and write the following formula for the total light attenuation:

$$I(x) = I(0) \exp(-\gamma x). \tag{3.3}$$

Equation (3.1) may be written in a different form:

$$I_t = I_i - I_r - I_a - I_s \tag{3.4a}$$

$$I_t/I_i = 1 - (I_r - I_a - I_s)/I_i = 1 - I_r/I_i - I_a/I_i - I_s/I_i. \tag{3.4b}$$

The incident collimated beam of irradiance I_i is divided in the following manner: part I_r/I_i is reflected, part I_s/I_i is scattered, part I_a/I_i is absorbed, and only part I_t/I_i is transmitted. We mentioned above that the ratio I_r/I_i is called *reflectance*.

The ratio I_t/I_i is sometimes called *transmittance* and it defines the fraction of the collimated beam that is transmitted, and one may similarly define the *absorbance* I_a/I_i and the *scattering ratio* I_s/I_i. Finally, the ratio $(I_r + I_s + I_a)/I_i$ is sometimes called the *attenuation ratio* and refers to the part that was actually lost due to reflection, scattering, and absorption.

Cases exist in which it is important to maximize the transmission of light through matter, as in optical fibers. In these cases, the manufacturers try to have low absorption and scattering coefficients and high transmittance at the wavelength of importance. In contrast, there are cases in which it is important to maximize the absorption in matter. For example, in laser surgery it is desirable to heat only a thin layer of tissue, in order to facilitate efficient vaporization (and cutting). The surgeon will thus use wavelengths (e.g., UV or IR) at which the absorbance in tissue is high. In other cases, it is intended to use laser light to heat large volumes of tissue (e.g., laser hyperthermia). In such cases, one often uses lasers (e.g., Nd:YAG) which give rise to significant scattering, due to a large scattering coefficient.

3.3.2 The Absorption of Laser Light in Materials

This section deals with the part of the radiation that is neither scattered nor reflected but absorbed. There are several ways in which such absorption can take place:

3.3.2.1 Fundamental Absorption

Light can be thought of either as a stream of particles called photons or as a succession of waves. Laser light can be viewed as a flow of photons moving in the same direction, all having the same energy and therefore the same "color." These are also waves of one frequency. When the laser beam passes through a material, the constituents of the material may absorb some energy. Often there is a minimum photon energy ϵ that can be absorbed. Photons of lower energy will pass through the material unhindered, retaining their energy. Photons of energy equal to or higher than ϵ will be absorbed. The fundamental absorption is related to the absorption of the constituents in the pure material itself. The fundamental absorption is a characteristic property of a pure material and varies from material to material. Normally, this is a strong absorption that results in a small extinction length. For example, silicate glasses are generally transparent in the visible part of the spectrum, but at a wavelength of about 330 nm in the UV the absorption increases sharply and transmission is low. The fundamental absorption of fused silica occurs at shorter wavelengths in the "deep" UV part of the optical spectrum (i.e., $\lambda < 300$ nm). Crystalline materials, organic materials, and inorganic materials each have a characteristic fundamental absorption. In most cases, the absorbed energy (i.e., the energy lost by the laser beam while passing through a sample) is converted into heat.

3.3.2.2 Absorption by Impurities

All materials are impure to some extent. Impurities are present intentionally or unintentionally and often absorb at some characteristic wavelength, where the host material does not absorb. Typically, when the concentration of these impurities is low (a few percent), the absorption is also low in comparison to the fundamental absorption. As previously mentioned, pure silica glass is transparent in the visible and is essentially colorless. The addition of 1% iron oxide gives rise to strong absorption in the visible, causing the glass to look deep green. Alternatively, addition of a small amount of copper gives rise to a green–blue color and nickel gives a yellow tint.

In the early days of endoscopy, the glass that was used to fabricate fiberoptic endoscopes contained small amounts of impurities. These gave rise to selective absorption and slight coloring of the glass. As a result, the picture obtained in the image bundle (see Section 5.5) had hues that were unnatural. With present-day endoscopes, the glasses used are very pure and the color rendition is highly accurate. In the power fibers discussed in Section 4.8, the host material must also be very pure. Impurities might absorb some of the laser light, the absorbed energy may be converted to heat, and the thermal energy developed may melt the fiber.

3.3.3 Luminescence and Fluorescence of Materials

The absorption of light in matter does not always result in generation of heat. There are cases in which absorbed light of one color excites the material and causes emission of light with a different color. This emission is known by the general term luminescence (Levi, 1968). If the luminescence appears only during the excitation period, it is called fluorescence. If it continues after the excitation, it is called delayed fluorescence or phosphorescence. These three types of luminescence processes are shown schematically in Fig. 3.4.

In general, when energy is absorbed in a material, electrons are excited from a lower level, called the ground state, to an upper level, called the excited state. Immediately after the excitation, some of the electrons fall back directly to the ground state. This process is fast (<10 nsec) and it gives rise to fluorescence. Some of the electrons fall back to an intermediate state, called a metastable state, from which they have to reach the ground state. In one case, the electrons must return to the excited state before they fall back to the ground state. This process is slower and gives rise to delayed fluorescence. Some of the electrons may fall directly from the metastable state to the ground state. This process, which is often slow, is termed phosphorescence. In the case of fluorescence or delayed fluorescence, the energy of the photon going in is often only slightly greater than the energy of the photon going out. Therefore the wavelength of the exciting light is only slightly less than that of the emitted light ($h\nu_{01} = h\nu_{in}$). In the case of phosphorescence, the energy of the emitted light is always much smaller than the ex-

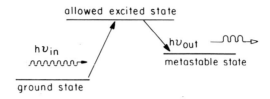

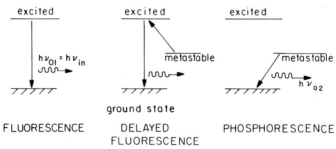

FIGURE 3.4   Luminescence: fluorescence, delayed fluorescence, phosphorescence.

citation energy ($h\nu_{02} < h\nu_{in}$). The emitted phosphorescence of light will therefore always be less energetic—that is, it will be shifted to the red part of the spectrum.

We will illustrate these phenomena with two examples:

(i) Fluorescent lamps: The glass tube in domestic fluorescent lamps is internally coated with a luminescent material. Electric discharge in the lamp causes the emission of invisible UV light; the material is exposed to the UV light and emits visible light. By covering the inside of the tube with different materials, one may obtain different emission wavelengths (i.e., different colors).

(ii) Television screen: The inner face of a television tube is coated with luminescent materials which emit visible light under excitation from an electron beam. In each luminescing "spot" on the TV screen, there is an array of different materials which emit red, blue, and green light. The combination of these emissions generates the desired color in that particular spot. When watching TV, the emission that we observe is the fluorescence. The emission that lasts a few seconds after the TV set is turned off is phosphorescence.

For a given process of luminescence (fluorescence or phosphorescence), one may measure two spectra: the emission spectrum and the excitation spectrum. The emission spectrum is obtained by exciting the luminescence with a particular wavelength. The intensity of the emission is then plotted as a function of wavelength. The excitation spectrum measures the emission at a particular wavelength (usually the wavelength of maximum emission). One may excite a sample with light of various wavelengths but equal irradiance. The excitation spectrum is obtained when the irradiance of the emission at the peak emission wavelength is plotted as a function of the incident light wavelength.

The case shown schematically in Fig. 3.4 involves only two or three energy levels. There are many other cases, especially those which are relevant to biology and medicine, in which a multitude of energy levels are involved. The excitation and emission processes often involve two "bands" of energy levels. As a result, both the excitation and the emission spectra are rather broad, with the emission band slightly shifted to the red, as shown in Fig. 3.5 (the corresponding spectra in Fig. 3.4, which are due to single levels, are narrow).

Many materials luminesce when excited by x-rays, electrons, UV, or a visible light source. The latter is called photoluminescence and refers to both fluorescence and phosphorescence. Luminescence that is due to the material itself is often called *intrinsic*. Different materials have different emission and excitation spectra, and the characteristic luminescence is one way of identifying materials. Luminescence may also result from certain impurities in the material. It is then called *extrinsic*. This is often a characteristic luminescence. For a given material, the presence of a particular luminescence confirms the presence of certain impurities. Moreover, the luminescence irradiance is often proportional to the concentration of impurities. By measuring it, one may (in principle) determine the impurity concentration.

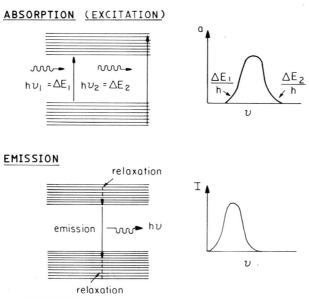

FIGURE 3.5 Luminescence: excitation and emission spectra.

Luminescence measurements usually require a flux of photons that have sufficient energy to induce photoluminescence. Energetic photons, in the blue or UV, are needed. In order to carry out the luminescence measurement, one needs a fairly high flux of these photons, that is, a light beam of sufficient irradiance. Lasers are convenient for this purpose both because they have high irradiance at a particular wavelength and because they can easily be directed onto a sample. For a given luminescence, one may choose a laser that emits at a wavelength that coincides with the peak of the excitation spectrum. These laser methods are often called laser-induced fluorescence (LIF) and may prove quite useful in some of the diagnostic techniques discussed below.

3.3.4 Material Processing by Laser Beams

In preparation for discussing the effects of laser beams on tissue, it is important first to understand, in general, how lasers are used for processing materials. This discussion is concerned only with the heating effects of the laser beam, which are caused by the absorbed laser energy.

Consider a laser beam with an area A (cm^2) and impinging on the surface of a material with power density (irradiance) I (W/cm^2). Let us assume that the beam is totally absorbed in a distance L (cm) below the surface. If the scattering effects are ignored, then power $P = IA$ (W) is absorbed in a cylinder whose base is area A and whose height is L. If the laser beam is applied for a period t, the energy deposited in the cylinder is $E = Pt$ (J) and the fluence (energy deposited in a unit area) is $F = E/A = It$ (J/cm^2). It is known that if energy E is absorbed in a piece of material of mass m and heat capacity c, its average change in temperature ΔT is given by Eq. (3.5b):

$$E = mc\ \Delta T \tag{3.5a}$$
$$E/mc = \Delta T. \tag{3.5b}$$

The validity of the equation is based on complete thermal isolation of the material. In real situations heat losses are present and the actual temperature rise is less. Also, it is assumed that there is no phase change (such as melting or evaporation, which consume energy). In this case, Pt can be substituted for E. If the density of the material is ρ, one can write $m = \rho LA$. Finally the average temperature rise ΔT is calculated as follows:

The total deposited energy is

$$E = Pt = (\rho LA)c\ \Delta T. \tag{3.6a}$$

The energy deposited in a unit volume is

$$\epsilon = Pt/LA = \rho c\ \Delta T. \tag{3.6b}$$

The temperature rise is

$$\Delta T = (Pt)/(\rho LAc) = (It)/(\rho Lc). \tag{3.6c}$$

If the initial temperature of the material is T_i, then the final temperature is $T_f = T_i + \Delta T$. If T_f is sufficiently high, the material may evaporate. We must then incorporate the latent heat of vaporization H. Again, if thermal losses are ignored, we obtain in this case the following formulas:

$$E = IAt = mc \, \Delta T + mH = \rho LA(c \, \Delta T + H) \qquad (3.7a)$$
$$F = E/A = It = \rho L[c \, \Delta T + H]. \qquad (3.7b)$$

One may use similar formulas to calculate the material removal rate. Let us use Eq. (3.7a) to calculate the energy dE needed to vaporize a layer of thickness dx during a time dt:

$$IA \, dt = dE = \rho cA \, dx \, \Delta T + \rho HA \, dx.$$

The vaporization rate is thus

$$u = dx/dt = I/(\rho c \, \Delta T + \rho H). \qquad (3.7c)$$

Material processing by laser beams is based on the temperature rise ΔT. For a given material, several parameters are at our disposal, including the wavelength of the laser, power, beam diameter, and duration of the exposure. The final ΔT is determined by the combination of these parameters.

(i) Pulsed operation: Consider a short laser pulse ($t < 1$ μsec) that is essentially totally absorbed in a material. Both the laser pulse and the absorption processes are optical processes that are fast. As mentioned earlier, however, thermal processes carry the heat away by radiation as well as by convection or conduction. The latter are rather slow (in terms of milliseconds); therefore, little heat is lost during the pulse if the volume of material in which the radiation is absorbed is isolated. The temperature rise ΔT can then be high.

(ii) CW operation: Consider the case of a continuous-wave (CW) laser interacting with a piece of material. Energy is constantly supplied to a cylindrical segment of the material. Most of the energy is absorbed, producing a temperature rise. If there were no other processes, Eq. (3.5b) would result in a boundless increase in T. As heat tries to build up in the cylinder, however, other processes take it away. Conduction of heat to neighboring pieces of the material, together with convection or radiation from the hot cylinder, contributes to heat loss. The system reaches equilibrium when the amount of energy absorbed per unit time is equal to the amount of energy lost per unit time. This equilibrium will occur at a certain temperature that can be calculated or measured. The equilibrium temperature T_{eq} is the average temperature of the cylinder during the exposure to the laser beam.

(iii) Wavelength: For each wavelength, there may be a different extinction length l. For most materials, it is possible to choose a laser beam that will penetrate deeply into the material or will be totally absorbed near its surface.

(iv) Power: A higher power level will produce a greater change in temperature over a given time.

(v) Beam diameter: One of the major advantages of the laser is that the Gaussian beam can be focused onto a small spot. The smallest spot size is roughly equal to the wavelength of the laser light itself [see Eq. (2.7)]. If a given power is concentrated in a smaller volume, ΔT will be higher, thus making it easier for the materials to be vaporized.

Based on these concepts, different types of interactions between lasers and materials can be considered and many of these interactions have been used for material processing.

3.4 LASER INTERACTION WITH TISSUE—PRINCIPLES

There is a wealth of information on the interaction of laser beams and tissue. Many aspects of this problem are described in conference proceedings and reviews that are fully dedicated to this issue (Berry and Harpole, 1986; Berns, 1988; Jacques, 1990, 1991, 1992; Preuss and Profio, 1990; Welch, 1985, 1991).

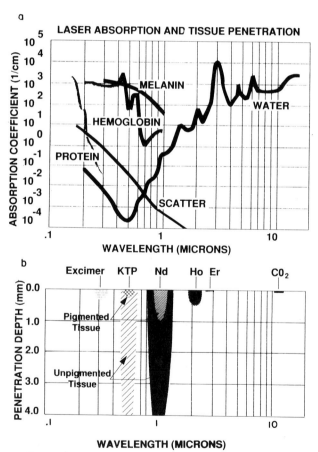

FIGURE 3.6 Spectral absorption in (a) water and (b) tissue. (Based on an illustration by Coherent Inc.)

3.4.1 Optical Properties of Tissue

Most tissues of the human body contain more than 70% water. Each tissue has its characteristic optical absorption spectra. For a first-order approximation one can state that the optical properties of tissue are similar to those of water. Many people have studied the spectral absorption of water at different wavelengths. The variation of the extinction coefficient of water with wavelength is shown in Fig. 3.6.

Both water and saline solution transmit well in the visible range and the absorption is high in the UV ($\lambda < 300$ nm) and the IR ($\lambda > 2$ μm). Tissue shows similar strong absorption in the UV and the IR. In blood, chromophores such as hemoglobin and bilirubin show strong absorption in the visible. Therefore, for a tissue that contains blood, the absorption in this range is dominated by the absorption in blood. Tissue may also contain other chromophores that absorb light in specific spectral regions, as also shown in Fig. 3.6a. The penetration depths for some of the important medical lasers are shown schematically in Fig. 3.6b. Figure 3.7 illustrates the penetration depth of some representative tissue (van Gemert and Welch, 1989).

Tissue highly absorbs UV and blue light but transmits red light, as mentioned in Section 3.2.1.1. It should be added that there is practically no scattering of laser light which is highly absorbed in tissue (e.g., from UV or IR lasers). Noticeable scattering is important only when dealing with visible or near IR lasers. This is shown schematically in Fig. 3.8.

Different types of tissue show markedly different optical and thermal properties. Some of these properties may depend, for example, on the water content of

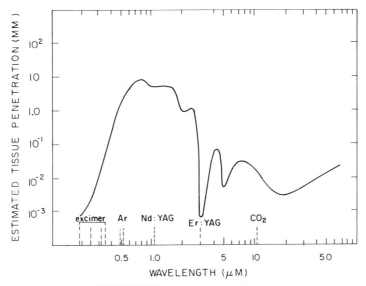

FIGURE 3.7 Penetration depth in tissue.

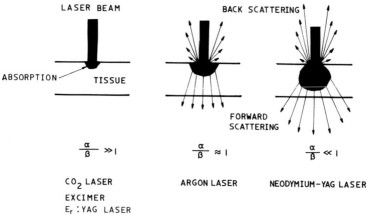

FIGURE 3.8 Absorption and scattering for the important lasers. α, absorption rate; β, scattering coefficient.

the tissue. During laser vaporization of tissue, the water content changes, causing the optical properties to vary.

3.4.2 Tissue Luminescence

Luminescence processes in materials have been discussed in Section 3.3.3. The emission or the excitation spectra can be measured when luminescence is excited. In many systems these spectra characteristically consist of narrow lines. Atoms or molecules that give rise to such spectra can be identified by these lines. Moreover, the emitted light depends on the concentration of the atoms. Generally, the higher the concentration, the higher the luminescence emission.

This section discusses the luminescence properties of tissue. The natural, intrinsic (e.g., endogenous) fluorescence of tissue itself, sometimes referred to as autofluorescence, is distinguished from the extrinsic (e.g., exogenous) fluorescence of reagents or dyes that are added to tissue or blood. The molecules that absorb the photons and cause fluorescence are called fluorophores.

Intrinsic (Endogenous) Fluorescence

Even in 18th century, luminescence from human tissue was being observed. It was noticed that, after exposing the skin to sunlight, it glowed in a dark room. Many intrinsic fluorophores, such as proteins, nucleic acids, and nucleotide co-enzymes, occur naturally. For proteins, the maximum of the excitation spectrum occurs at 280 nm and the emission spectra have maxima at 320–350 nm. The nucleotide coenzyme NADH has an excitation spectrum with a maximum at 340 nm and emission maxima at 450–470 nm. Fluorophores in collagen and elastin show an emission peak at 380 nm, and melanin has a broad emission peak at 540 nm.

Extrinsic (Exogenous) Fluorescence

If fluorescence in biological tissue is excited by a UV light source, the emission will usually extend over a wide range of wavelengths. This is a result of simultaneous emission from the various intrinsic fluorophores in the tissue. The analysis of autofluorescence from tissue is therefore rather complex. The diagnostic capabilities of fluorescence spectroscopy may be enhanced by incorporating extrinsic fluorophores in tissue. Numerous fluorophores can be utilized; several are discussed in Section 3.5.

3.4.3 Photothermal Effects

Biological tissues are complex systems, and this makes the interaction between laser beams and biological systems quite complicated. There is a tendency for physicists (this author included) to oversimplify these problems. If it is assumed that tissue consists entirely of water, simplifications can be made when studying laser–tissue interactions. These are given in the following and better approximations are given in the literature (Bonner *et al.*, 1986; Thomsen, 1991). A more general description is given later on.

Consider a laser beam of a certain wavelength λ that impinges on water and is absorbed. The absorbed laser energy causes heating of a certain volume of water, resulting in a temperature increase. In this section we present nonablative and ablative heating effects.

3.4.3.1 Heating—Nonablative Effects

In Section 3.3.4 we discussed heating a sample by a laser beam. Let us use Eq. (3.6) to calculate the temperature rise ΔT in a sample after exposure to Nd : YAG laser light.

EXAMPLE —*Heating of Water*: Assume that an Nd : YAG laser beam of area $A = 1$ mm$^2 = 0.01$ cm^2 impinges on water at room temperature, $T_i = 20°C$. We also assume that the beam is totally absorbed in a depth $L \approx 1$ cm (and therefore the energy is absorbed in a volume of $0.01 \times 1 = 0.01$ cm^3). If the laser power $P = 1$ W is applied for $t = 2$ sec, the energy delivered is $Pt = 2$ J ≈ 0.5 cal. The density of water is $\rho = 1$ g/cm^3 and its specific heat is $c = 1$ cal/g · deg. By substituting these values in Eq. (3.6), ΔT can be calculated:

$$\Delta T = 0.5 \times 1/0.01 = 50°C.$$

The volume exposed to the laser beam is thus heated to $T_f = T_i + \Delta T = 70°C$. Although this is an approximation, it demonstrates the heating principle. It must be emphasized that scattering effects, which will increase the heated volume (and decrease the average temperature), are neglected in this example. Also, heat losses are neglected.

3.4.3.2 Vaporization—Ablative Effects

Let us consider again the heating of water. As mentioned, water has a density of $\rho = 1$ g/cm^3 and a specific heat of $c = 1$ cal (g °C)$^{-1}$. Whereas 1 cal = 4.2 J is needed to heat 1 g of water by 1°C, approximately 265 J = 63 cal are needed to raise the temperature of 1 g of water from body temperature T_i = 37°C to T_a = 100°C (ΔT = 63°). In addition, the latent heat of vaporization of water is H = 540 cal/g and 540 cal = 540 × 4.2 J = 2270 J are required to convert 1 g of water at 100°C from liquid to vapor. Therefore about 2500 J are needed to raise 1 cm³ of water from body temperature (37°C) to 100°C and to vaporize the water at this temperature.

Let us again consider a laser beam of area A that is absorbed within a length L. If the laser beam of energy Pt is absorbed in a volume V = AL of water[6], the minimum energy needed to vaporize the water can be calculated according to Eqs. (3.6) and (3.7). The threshold energy for vaporization is

$$E_{th} = Pt = 2500AL \text{ J}. \tag{3.8a}$$

The threshold energy density (per unit volume) for vaporization is

$$\epsilon_{th} = Pt/LA = 2500 \text{ J}. \tag{3.8b}$$

The threshold fluence (energy per unit area) for vaporization is

$$F_{th} = It = 2500L \text{ J/cm}^2. \tag{3.8c}$$

These equations give the energy threshold E_{th} or the fluence threshold F_{th} for vaporization.

EXAMPLE — *Vaporization of Water*: A CO_2 laser beam of spot size A = 1 mm² = 0.01 cm² impinges on water. The penetration depth is α = 1000 cm⁻¹ (see Fig. 3.6a), and we assume $L = 4/\alpha$ = 0.004 cm. The beam is fully absorbed in a volume of $LA = 4 \times 10^{-5}$ cm³. The following parameters may thus be calculated:

- Ablation threshold: If A and L are known, the minimum energy E_{th} needed to vaporize the water in a disk of area A may be calculated:

$$E_{th} = 2500 \times 10^{-2} \times 0.04 = 0.1 \text{ J} = 100 \text{ mJ}.$$

The minimum fluence needed for the ablation threshold is 100 mJ/mm² or 10 J/cm².

- Time: Let us consider a laser beam of constant power P = 10 W. In order to vaporize water, this laser beam must be operated for t sec. By writing $Pt = 0.1$ J, we see that $t = 0.01$ sec = 10 msec. An exposure of this length is needed to reach the ablation threshold.
- Removal rate: When operating the laser for 10 msec, a disk with a thickness of 0.004 cm = 40 μm was vaporized. One can therefore loosely infer

---

[6] Some authors use for L the penetration depth $1/\alpha$, some use the extinction length $4.6/\alpha$, and others use the value $2/\alpha$. Therefore the results cited in the literature reflect these differences. In this section we use $L = 4/\alpha$.

TABLE 3.1 **Water Vaporization by CO_2 Laser**

Heating of 1 ml of water from 37 to 100°C   63 cal
Vaporization of 1 ml of water at 100°C   541 cal
Total vaporization energy $\approx$ 600 cal $\approx$ 2500 J
Absorption coefficient $\alpha = 1000$ cm^{-1}
Extinction length $L = 4/\alpha = 0.004$ cm

For a beam of area $A = 1$ cm^2 the volume vaporized is $V \approx 0.004$ cm^3
An energy of $2500 \times 0.004 \approx 10$ J is needed for this vaporization
Comment: For a beam of area $A = 1$ mm^2 we obtain an energy of 100 mJ
The ablation threshold for water is ≈ 10 J/cm^2
The removal rate for water is ≈ 0.4 mm^3/J

that the "thickness" removal rate of water is 40 μm/10 msec $= 0.4$ cm/sec. Alternatively, one may state that a volume of 4×10^{-5} cm^3 was removed by a pulse of 100 mJ and therefore the "bulk" removal rate is 4×10^{-5} cm^3/100 mJ $= 4 \times 10^{-4}$ cm^3/J. Some of these values are given in Table 3.1.

3.4.4 Photochemical Effects

Almost 100 years ago, it was discovered that light-absorbing chemicals can cause photoreactions in biological systems. In 1900, Raab discovered that paramecia, sensitized by acridine dye, will perish when exposed to light in the presence of oxygen. This so-called photodynamic effect was also studied by Von Tappeiner at the same year. Von Tappeiner, in conjunction with Jesionek, suggested in 1903 that this effect might be used to treat skin diseases such as herpes or psoriasis. In clinical studies they found that topical application of aqueous solutions of eosin or fluorescein to skin tumors, followed by exposure to light, resulted in regression of some tumors. One of the photosensitizers that was discovered in the second half of the 19th century was hematoporphyrin, a product that is prepared by sulfuric acid treatment of hemoglobin. Since the 1800s, it has been known that hematoporphyrin is selectively retained in tumors. Hematoporphyrin has the ability to mediate photochemical reactions when illuminated by light of an appropriate wavelength. In the 1900s, this knowledge led to a number of efforts to utilize it for both detecting and treating malignant tumors. It was then found, in the early 1960s, that a hematoporphyrin derivative (HPD) has a greater affinity for tumors and that a characteristic red fluorescence is emitted when HPD is exposed to UV light. Lipson, in 1960, first used these properties for endoscopic diagnosis of cancer in the bronchus and later (1964) in the esophagus and the cervix. In 1967, Lipson reported on the first successful use of HPD for the photodynamic therapy of breast cancer. In the early 1970s, several groups reestablished efforts to use HPD and red lasers for photodynamic therapy of cancer. Some of the important developments in this field have been discussed in review articles (Dougherty, 1989; Marcus, 1992) and books (Kessel, 1990; Henderson and Dougherty, 1992).

3.4.5 Photomechanical and Photoablative Effects

When absorbed by tissue, laser pulses of short duration (e.g., <1 μsec) and relatively high power density generate stress waves. Many mechanisms that are responsible for these waves depend on the nature of the pulses and tissue. In this section, we briefly describe the generation of plasma and shock waves in transparent media and the photoacoustic effects in opaque materials (Deutch, 1988). Section 3.7.3 discusses their applications in therapy and surgery.

3.4.5.1 Laser-Produced Plasma and Shock Waves—Transparent Materials

Short and energetic laser pulses which are focused inside a *transparent* material (e.g., water) may give rise to a phenomenon called laser-induced breakdown. In the focused laser beam region, free electrons are released because of thermionic emission or multiphoton effects. The number of electrons increases rapidly and a plasma is produced. The (physical) plasma is at a very high temperature and is composed of a large number of electrons and positive ions. As the plasma continues to grow during the laser pulse, it absorbs the laser light that impinges on the focal spot and finally breakdown occurs. This dielectric breakdown is accompanied by high-intensity light emission. Such laser-induced breakdown was observed when Nd : YAG laser pulses of duration <10 nsec and energy >20 mJ were focused in water.

At the end of the laser pulse the plasma expands. A shock wave results which initially propagates at supersonic speeds and may penetrate tens of micrometers in the liquid.

3.4.5.2 Photoacoustic Effects—Opaque Materials

Stress waves may be generated in *absorbing* samples at laser power densities much lower than those needed for dielectric breakdown. These waves are caused by two effects:

(i) *The thermoelastic effect*: The absorbed laser beam results in sample heating, which causes expansion and gives rise to large stresses. Rapid heating by a strongly absorbed laser pulse generates stress waves that propagate from the heated area inside the sample.

(ii) *Ablative recoil*: The absorbed energy causes ablation and the recoil of the removed material imparts momentum to the sample surface. Although the mass of material removed is generally small, the ablation velocities may be quite high, causing a large momentum. This leads to the generation of strong stress waves that propagate in the sample.

Both the thermoelastic effect and the ablative recoil may modify the tissue behavior. Differences are evident if the sample is in air or immersed in transparent liquid.

3.5 LASER-ASSISTED DIAGNOSTICS—PRINCIPLES

3.5.1 Spectrophotometric Methods

When laser light of a certain wavelength is incident on tissue, it is possible to measure the reflectance (i.e., the ratio between reflected light and incident light). This process is performed for a variety of wavelengths and results in a spectral measurement known as the reflectance spectrum. Reflection spectrophotometry is complementary absorption spectrophotometry. In biomedicine, reflectance spectrophotometry may be performed on the skin using tunable dye lasers in the visible range. This is a noninvasive technique that provides information on pigment levels in tissue and, in particular, melanin and hemoglobin in the skin. Visible light penetrates 1–2 mm into the skin and therefore the reflectance measurements may supply information about the hemoglobin content in tissue and its degree of oxygenation. Reflectance spectrophotometry may also provide information about malignant tissues. For example, some laser therapy methods are based on damage to blood vessels in tumors, which may be revealed by spectrophotometric methods. These methods are easily adaptable to use with optical fibers, as will be discussed in Chapter 7.

3.5.2 Luminescence Methods

Both the emission and the absorption spectra in biological systems are generally broad and often featureless. Nevertheless, luminescence methods (e.g., laser-induced fluorescence) can be used as a diagnostic tool in biology and medicine. In some cases, the intrinsic luminescence (excitation or emission) spectra differ between healthy and unhealthy tissue. In other cases, a fluorophore injected into the body accumulates only in one type of tissue. Under excitation, only this tissue shows extrinsic luminescence. Several specific examples of the use of luminescence in preclinical or clinical trials are discussed in the following. Emphasis is placed on the luminescence methods that are used endoscopically.

An additional aspect of luminescence is its temporal behavior. The luminescence produced by excitation by a short pulse will decay in a period called the decay time constant. As discussed earlier, fluorescence decay is rather short, whereas phosphorescence decay is fairly long. The time constant is also a characteristic of the luminescent species.

In this section we discuss the use of these techniques in oncology and cardiology. Similar diagnostic methods have been tried in other areas. For example, laser light absorption, scattering, and luminescence were used to detect caries in human teeth (Alfano *et al.*, 1984)

3.5.2.1 Diagnosis of Cancer

Several groups (Andersson-Engels *et al.*, 1989, 1990) have found that the autofluorescence (i.e., endogenous fluorescence) of malignant tumors is different from that of normal tissue. This autofluorescence is shown schematically in Fig. 3.9 and may be used to both detect the presence of tumors and identify the demarcation lines between tumor and normal tissue.

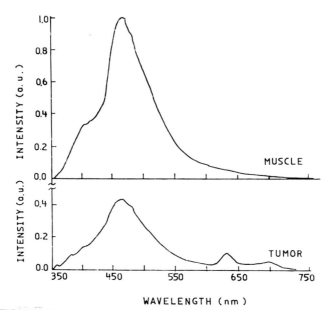

FIGURE 3.9 Autofluorescence in tissue: malignant tumor and healthy muscle.

When certain compounds, such as porphyrins or other agents, are injected into the body, they eventually concentrate in malignant tumors. These nontoxic compounds fluoresce under UV or violet excitation, and malignant tumors are therefore detectable by their emissions (e.g., exogenous fluorescence) of orange and red light. The fluorescence of HPD is mostly in the red, with two prominent emission peaks at 630 and 690 nm, as shown in Fig. 3.10a. The excitation spectrum of the 630-nm emission line peaks at around 400 nm. The fluorescence efficiency is defined as the ratio of the emitted power to the excitation power. In the case of HPD, the fluorescence efficiency is fairly low (a few percent). It is desirable to use a laser beam for excitation because it emits enough power to cause visible fluorescence. A krypton ion laser emitting 0.2–0.3 W at 413 nm is the most suitable for this application. The characteristic red luminescence has been used to identify malignant tumors.

In principle, the exogenous fluorescence of HPD may be used for fluorescence imaging. The tumor and the surrounding normal tissue is illuminated with laser light at ≈410 nm, and the whole area is then imaged using a high-quality video camera. A special filter that transmits only red light at around 630 nm is placed in front of the camera lens so that only the red-emitting tumor is imaged. In practice, there were significant difficulties due to the autofluorescence of tissue, both in the blue (i.e., ≈470 nm) and in the red (i.e., ≈600 nm) that prevented clear imaging of tumors. There have been recent attempts to solve the problem using image-processing techniques. For example Anderson-Engels et al. (1990) imaged the tissue three times, using filters that transmitted the wavelengths 630 nm, 600 nm, and 470 nm. For each point on the image they thus obtained three values

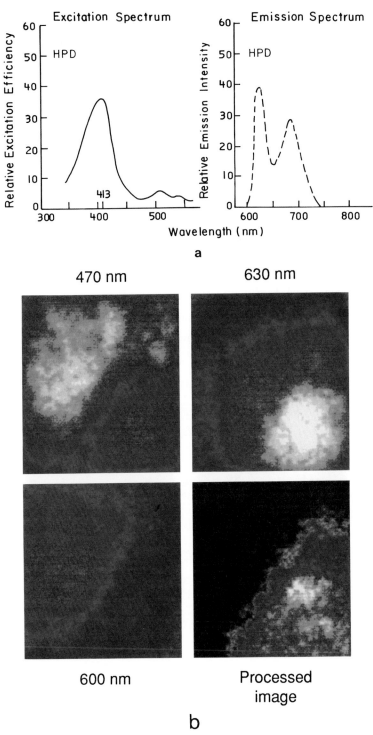

FIGURE 3.10 (a) Excitation and emission spectra of HPD in tumor. (b) Fluorescence imaging of brain tumor in rats recorded at three wavelengths and the computer-generated image (Courtesy of K. Svaanberg.)

of emission intensity, I(630), I(600), and I(470). A computer was then used to calculate at each point the relative value $L = [I(630) - I(600)]/I(470)$ and generate an image showing L at each point. This image shows a clearer demarcation between tumor and normal tissue. Results obtained in rats inoculated with gliomas in the brain are shown in Fig. 3.10b. The image obtained through the red (630 nm) filter shows the tumor, but the edges are diffuse. A much clearer demarcation of the tumor is obtained with the computer-generated image.

The UV light used for excitation is highly absorbed in the tumor, and therefore fluorescence is emitted from its outer surface. This diagnostic method may therefore detect superficial tumors, such as lung tumors at an early stage. This is important when the mass of the tumor is rather small and cannot be detected by other methods such as ordinary x-ray imaging, computed tomography (CT), or magnetic resonance imaging (MRI). When tissue is excited by short UV pulses, it fluoresces. The fluorescence decay time is different for normal and malignant tissues which contain HPD. The temporal behavior of the fluorescence may also be used to diagnose cancer.

3.5.2.2 Detection of Plaque in Cardiology

In endoscopic cardiology and cardiovascular surgery, it is necessary to distinguish between plaque and normal tissue. Luminescence techniques may also be

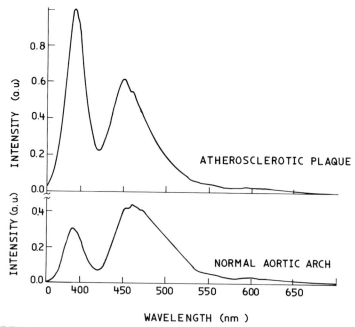

FIGURE 3.11   Autofluorescence in tissue: plaque and normal artery wall.

useful here. There are several reports which indicate that autofluorescence from plaque has a different luminescence from normal tissue. Studies also show (Kittrell *et al.*, 1985; Leon *et al.*, 1988) that both the emission spectrum and the excitation spectra are different. For example, the work of Kittrell *et al.* proved the existence of a peak in the plaque emission spectrum that is absent in normal tissue. Tissue under study may be illuminated, for example, by specific UV light (which gives the characteristic luminescence). The emission spectrum, as shown in Fig. 3.11, is then analyzed and the results used to distinguish between healthy tissue and plaque.

Reports have also shown that plaque behaves like a malignant tumor in the sense that it selectively retains some chromophores. If illuminated by UV light, photosensitized plaque fluoresces with a characteristic emission which is absent in the autofluorescence of normal tissue. The luminescent agent must be nontoxic, be selectively retained in plaque for a few hours, and have a characteristic emission. Preferential accumulation of tetracycline in plaque was observed in the 1960s, and more recently HPD was tried as an exogenous chromophore (Spears *et al.*, 1983).

3.6 APPLICATIONS OF LASERS IN DIAGNOSIS AND IMAGING—ADVANCES

3.6.1 Raman Spectroscopy

The reflection, absorption, and laser-induced fluorescence (LIF) methods are based on electronic transitions. They indicate the biochemical composition of tissue and have been used to distinguish between healthy and diseased tissue. There are other methods which are based on the vibrational transitions in molecules. In particular, molecular bonds can be detected by Raman spectroscopy, and this can be used to determine the relative concentration of specific biomolecules in tissue. For example, the Raman spectrum of soft atherosclerotic plaque is similar to that of cholesterol and is different from that of calcified plaque or of the normal arterial wall. Like laser-induced fluorescence, Raman spectroscopy can be performed on some tissue and it lends itself to biochemical imaging of the tissue. Moreover, Raman spectroscopy can be performed through optical fibers, so the imaging could be carried out through an endoscope (see Chapter 6).

3.6.2 Time-Dependent Spectroscopy and Imaging

In the ultraviolet and infrared spectral ranges, the absorption coefficient α is much larger than the scattering coefficient β. In these ranges there is little scattering by tissue and the light absorption follows the Beer–Lambert law [Eq. (3.2)].

In the spectral range 0.5–1.3 μm the absorption coefficient (e.g., $\alpha \approx 10$ cm^{-1}) is much smaller than the scattering coefficient (e.g., $\beta \approx 10$–100 cm^{-1}). When light in this far-red and near-infrared range impinges on soft tissue, it will penetrate deeply because of the low absorption and high scattering. This spectral range is therefore called the optical "window" of tissue. For light in this spectral range one may carry out two types of diagnostic measurements on the tissue: (1) spectroscopy, measurements of the reemitted light as a function of wavelength, and (2) imaging, measurements of the spatial distribution of absorbing or scattering structures in the tissue. Optical spectroscopy and optical imaging measurements have been carried out on tissue using ordinary light sources and CW lasers. The information obtained from these measurements was severely limited by the strong scattering of light. The situation was changed dramatically with the introduction of time-dependent (also called time-resolved) methods.

Photons emitted by a light source are incident on a tissue sample and are detected by a detector situated some distance away. The time taken by the photons to reach the detector (called "time of flight") is proportional to the path length of the photons. A few photons (called "ballistic" photons) pass through the tissue without scattering, and their time of flight is the shortest. For most of the photons the path length is much longer and they are delayed. Some of the lasers mentioned in Chapter 2 emit near-infrared light in pulses shorter than 1 nsec (e.g., mode-locked dye lasers or diode lasers). There are light detectors that can detect such fast pulses. With these one could measure directly the time of flight. When a short laser pulse is used as an input pulse that is incident on tissue, the output pulse will be broadened (in time) because of the different path lengths of the photons. Various techniques (e.g., "time gating") have been developed to determine the time of flight of the ballistic photons. In other words, one could eliminate the effects of the scattered photons and study only the minimally scattered photons.

It was found (Benaron, in Caro and Choy, 1992; Hebden *et al.*, 1991; Wilson *et al.*, 1992) that one could use these methods for optical spectroscopy and for imaging of tissue. In spectroscopy both reflectance and absorption measurements can be performed and can be used for quantitative measurements on tissue, such as the hemoglobin oxygen saturation. The optical imaging techniques used are similar, in a sense, to x-ray imaging techniques. Ordinary radiography is a direct method in which x-rays that are transmitted through a sample generate a two-dimensional image on a screen. Computed tomography is an indirect method in which narrow beams of rays are transmitted through a sample and a three-dimensional image is reconstructed mathematically. In optical imaging the direct method is called direct projection or direct transillumination. The indirect methods are called optical tomography. Both methods were used first *in vivo* to detect objects in a turbid medium and then *in vivo* to try to detect tumors (e.g., breast tumors) in the body.

Time-dependent measurements can be performed via optical fibers. There is therefore great interest in carrying out time-resolved spectroscopic and imaging measurements using laser catheters or laser endoscopes.

3.7 LASER SURGERY AND THERAPY—PRINCIPLES

Various effects, depending on the temperature increase, are observed when laser light is absorbed in tissue and converted into thermal energy. These effects have been fully discussed in conference proceedings (Joffe *et al.*, 1989; Joffe and Atsumi, 1990) and review papers (Deutch, 1988). The interactions between a laser beam and tissue are illustrated by a general map that is similar to the one presented by several authors (Boulnois, 1986) and is shown in Fig. 3.12. In this figure, the power density of the beam (W/cm^2) and the interaction time with tissue (sec) are both plotted on a logarithmic scale. The diagonal lines denote constant fluence (J/cm^2). Various effects are indicated: (i) photochemical effects at low power densities and long interaction times, (ii) nonablative heating effects (e.g., hyperthermia, coagulation, and welding) and ablative heating effects (e.g., surgery) at higher power densities and shorter interaction times, and (iii) photoablative and photodisruptive effects at very high power densities and very short interaction times. These effects are discussed below.

3.7.1 Photothermal Mechanisms

3.7.1.1 Laser Hyperthermia

Many studies have shown that at certain elevated temperatures, malignant tissue is damaged with relatively little effect on normal tissue. Furthermore, con-

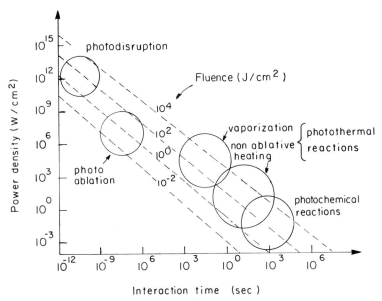

FIGURE 3.12 General graph of laser interaction with tissue for various power levels P (W/cm^2) and interaction times τ (sec).

trolled heating of tumor tissue to therapeutic temperature levels may lead to regression and complete tumor remission. The minimum tumor temperature needed to achieve this effect is 42.5–43.0°C and must be maintained for 20–60 min to be effective. The effectiveness of this modality depends, to a large extent, on the ability to establish and maintain uniformity of the tumor's temperature. Some of the early work on hyperthermia involved systemic (whole body) heating; however, this modality is relatively risky for patients. A preferable modality is localized hyperthermia, where great effort is made to restrict the temperature elevation to the target tissue with minimal thermal damage to neighboring normal tissue.

For localized and large tumors, an electromagnetic field that induces heating, such as a microwave or radio-frequency field, or ultrasound heating may be used. Lasers offer an alternative for smaller tumors. Laser energy that is not well absorbed by tissue (e.g., Nd : YAG laser light in the near IR) will penetrate deeply into the tissue, causing a controlled rise in the temperature. The near-IR light is easily delivered through silica-based glass fibers. Fibers may be a useful tool for hyperthermia treatment of tumors inside the body (Steger *et al.*, 1989), as discussed in Section 9.7.5.

3.7.1.2 Laser Coagulation

Laser energy absorbed in blood causes coagulation at a temperature of 60°C. This effect, observed in the early days of laser surgery, was used for laser-induced hemostasis of bleeding ulcers, hemorrhoids, and so forth. Two types of lasers have been commonly used: an Ar ion laser, whose wavelength is strongly absorbed in blood, and an Nd : YAG laser, whose wavelength is poorly absorbed. The Nd : YAG laser heats the tissue, resulting in contraction which squeezes the blood vessels shut. An Ar laser does the same but affects a much smaller volume of tissue.

3.7.1.3 Laser Welding of Tissue

A coagulation bond may be formed between tissues when the temperature is elevated (e.g., by electrical heating). As a replacement for suturing, local heating was tried as a method for connecting two segments of soft tissue. Two pieces of tissue placed next to each other will also be connected when a laser beam is applied and the interface area is heated. This is called laser welding of tissue, and the first experiments in this technique were reported in the mid-1960s (Yahr and Strully, 1966). Sutureless welding by lasers has several advantages: it is inherently sterile and nontactile, no foreign materials, such as suture material, are utilized, and it may also save time. The welded surface is smooth and grows without restriction. At the same time there are a few disadvantages: the welding is not well controlled and the *initial* strength of the weld is lower than that of sutures.

The mechanism by which soft tissues are glued to each other is under investigation. Although some researchers emphasize the photochemical aspects of the process, there are strong indications that thermal effects are crucial. One possible explanation of the process is related to the coagulation of proteins. Biological cells are essentially little sacks filled with water, proteins, fat, and carbohydrates. When heated by a laser beam, disruption of the cells may result, followed by an overflow of a watery protein solution. While the proteins denature and adhere together, this solution will coagulate on heating. This is illustrated by what happens when one boils an egg, where coagulation and bonding effects caused by heat are obvious. The coagulated proteins probably serve as the biological glue that holds together the pieces of soft tissue (Bass *et al.*, 1992).

Several lasers have been used for laser welding, such as Ar ion and Nd:YAG. In some cases the laser heats the tissue itself. In other cases "natural adhesives" are used. These are protein-type solutions, such as blood or human fibrinogen, which are applied on the tissue. Under the thermal effects of laser energy these materials act as adhesives that reinforce the welded tissue. The highly absorbed CO_2 laser beam may be better for use with these adhesives.

A few of the potential applications of laser welding are described:

(i) Anastomosis: Blood vessels can be connected by laser welding. Until now, experiments connected veins or arteries by two or three "stay" sutures which held the edges together. A laser beam was then applied around the circumference, leading to laser welding. Not only are there indications that healing is faster with the latter method, but also *after healing* the laser welding did not result in aneurysm or stenosis, common complications of conventional suture methods.

(ii) Wound closure: The edges of a wound can be held together by a clamp or a few sutures. Heating by a laser beam seals the wound.

(iii) Nerve attachment: Some of the early experiments were aimed at welding severed nerve cells. Several lasers have been tried for nerve anastomosis, such as Ar ion, Nd:YAG, and CO_2. Some experiments were performed in air, others under saline solution. The temperature rise was measured and found to vary from 45 to 70°C. These processes are not yet fully understood and no agreement has been reached on the most suitable modality for each of these applications. Yet there are some indications that nerve attachment is possible, if the nerves are carefully cut before the welding. If the cutting is performed with an excimer laser, the two ends to be welded are not damaged thermally, leading to better results.

(iv) Delicate structures: Laser welding has also been tried for delicate structures such as fallopian tubes, skin, ureter, and microvessels. Laser welding offers watertightedness that is important in the bile duct, the intestines, blood vessels, and the ureter.

Laser welding is of special interest here because it is compatible with fiber-optic techniques. Laser welding will be used in areas where access is limited, where precision application is needed, and where the initial weakness of the weld is not a problem.

3.7.1.4 Coagulation Necrosis

Tissue necrosis (death) occurs when tissue temperature is raised to values higher than 45°C for a sufficiently long time. This is also explained in Section 3.8.2. This process has been used clinically for the destruction of large volumes of tissue. For example, a deep penetrating laser beam has been used to raise the temperature of large tumors to temperatures higher than 50°C for a few minutes. Coagulation necrosis of tissue occurred, and it was used as a palliative treatment both for malignant and benign tumors.

Strictly speaking, this is also a hyperthermia treatment. However, although the temperature is raised to values higher than discussed in Section 3.7.1.1, the effects are distinctly different.

3.7.1.5 Laser Vaporization

Although tissue is not water, it does consist mostly of water. A zero-order approximation is based on a simple model in which water is substituted for tissue. For a first-order approximation, one must calculate the ablation rate, in addition to the temperature rise in tissue, as a result of laser beam absorption. The exact behavior of tissue under the action of laser light is rather complex and beyond the scope of this book (Jacques, 1990, 1991, 1992).

If tissue does not have strong absorption, one may add chromophores that do absorb. For example, in an effort to destroy atherosclerotic plaque, some of the compounds that are selectively retained in plaque may be used. There have been reports of using carotene (Prince *et al.*, 1988) for this purpose.

3.7.2 Photochemical Mechanisms

In this section, we discuss therapy due to photochemical effects and in particular the principles of photodynamic therapy (PDT), including sensitizers, light sources, and potential clinical applications. Chapter 9 emphasizes the use of laser–fiber systems for endoscopic PDT.

3.7.2.1 Photodynamic Therapy

Photodynamic therapy (PDT) is based on a sensitizer that has the following properties:

- (i) The sensitizer is nontoxic and may be given intravenously.
- (ii) The sensitizer is either selectively taken up or selectively retained in malignant tumors.
- (iii) The concentrated sensitizer can be detected by its characteristic fluorescence emission, thus facilitating the detection of the tumor.
- (iv) The concentrated sensitizer can be activated by light of a particular wavelength that penetrates deeply (>5 mm) into the tumor.
- (v) The photoactivated sensitizer causes destruction of the tumor, including large tumors.

Although the sensitizers that have been developed up to now do not meet all these requirements, some come close and are potentially useful. In a photosensitized reaction, light is absorbed by a molecule (sensitizer), which leads to excitation by energy or electron transfer of a host molecule (substrate). In the case of photodynamic therapy, researchers have tried to find photosensitizers that are retained by tumors and for which the excited substrate is toxic to the tumor cells. Most of the work in this area was done on hematoporphyrin derivative and a related compound, dihematoporphyrin ether (DHE), and the principles of their operation are discussed here.[7]

Let us assume that the sensitizer S is excited by a photon of energy $h\nu$ from its ground state ^0S to an excited[8] singlet state $^1S^*$ (the electron spins are paired). When a sensitizer molecule is excited, the lifetime in this excited state is very short ($10^{-9}-10^{-6}$ sec) and two reactions are possible. The molecule may return to the singlet ground (lowest) state by emission of red characteristic light at wavelengths 600–700 nm. This is the characteristic emission which is used for cancer diagnosis (see Fig. 3.10). Alternatively the electron changes its spin state (intersystem crossing) and there is a conversion to an excited triplet state $^3S^*$ (parallel spins). This triplet state has a longer lifetime (10^{-3} sec) and has more time to interact with the surroundings. There are again two types of interactions: type I, which involves direct interaction of $^3S^*$ with the substrate, and type II, an energy transfer from the excited triplet state sensitizer to molecular oxygen ($^3O_2$) in the cells to produce singlet oxygen ($^1O_2$). The singlet oxygen has a lifetime of tens of μsec and therefore interacts only with materials in its vicinity. Singlet oxygen is electrophilic and efficiently produces oxidized forms of biomolecules. This is a reactive form of oxygen that interacts easily with other molecules (e.g., the substrate) to form oxidized products which are cytotoxic. In clinical work, the sensitizer is retained in tumor cells. When exposed to red light at 630 nm, the singlet oxygen is formed within the tumors, where a cytotoxic effect occurs.

The HPD PDT mechanism is shown in the schematic representation below:

$$S + h\nu \rightarrow {}^1S^* \rightarrow {}^3S^* \text{ intersystem crossing}$$
$$^3S^* + {}^3O_2 \rightarrow S + {}^1O_2 \text{ energy transfer}$$
$$^1O_2 + A \rightarrow AO \text{ oxidation}$$

3.7.2.2 Sensitizers

HPD or DHE can be photosensitized by various wavelengths in the visible spectrum. The excitation process is a "threshold" process; that is, there is a minimum power density that is needed to activate the sensitizer. In practice, the outer surface of the tumor is exposed to the excitation light at a power density level higher than the threshold and photoactivates it. The power density decreases inside the tumor. At a certain depth, it reaches a value below the threshold, where HPD is no longer activated. The excitation efficiency is highest for blue or violet. But at these wavelengths, absorbance is high and therefore the light is absorbed within a thin surface layer and does not penetrate deep into the tissue. The excitation efficiency for red light at 630 nm is low but the absorbance is also low. If

---

[7] In this book we will not distinguish between HPD and DHE (also known by a trade name PHOTO-FRIN); when HPD is mentioned, it may well refer to DHE.

[8] An excited state is denoted by *.

the red light power density at the tumor surface is sufficiently high, a depth of several millimeters it may reach values higher than the threshold for photoactivation. In the case of large tumors, optical fiber tips may be inserted inside the tumor for better and more uniform illumination of the bulk (called interstitial treatment). Red light at 630 nm is again chosen because it penetrates much deeper into the tumor. This is why HPD is most commonly activated by red light.

New sensitizers (e.g., chlorins, phthalocyanines, and purpurins) are being sought which could be activated by longer wavelengths, causing light to penetrate deeper into the tumor. The lasers mentioned below could be used to photoactivate these sensitizers in the far red.

3.7.2.3 Light Sources

As discussed, the optimal wavelength for the photoactivation of HPD in tumors is 630 nm. Although laser sources are not necessary for the excitation of HPD, they are highly useful because of their high irradiance, wavelength selectivity, and the need to couple the light energy into optical fibers. These fibers are used to deliver the laser energy from the laser to the irradiated tumor and they are fully discussed in Chapters 4 and 8. It is estimated that approximately $10-50$ mW/cm^2 must be applied to trigger the reaction. Both CW and pulsed lasers have been used for HPD–PDT:

(i) A CW dye laser which is itself pumped by an Ar ion laser: This is a tunable laser whose emission wavelength can be varied over a wide range in the visible, but the total CW power output is somewhat limited.

(ii) A pulsed dye lasers that is pumped, for example, by an excimer laser: This laser emits a train of short and high-power pulses

(iii) A pulsed Au vapor laser: Although it emits at a constant wavelength (628 nm), it maintains a higher average power output than the argon-pumped dye laser.

It has not yet been determined if the effects of the pulsed lasers are identical to those of the CW ones. There has also been a rapid development of high-irradiance semiconductor lasers which can continuously emit several watts in the near IR. These lasers are being widely used for laser printers, compact disc players, and so forth. There has been an effort to extend the emission wavelength of these lasers to the red part of the spectrum (see Section 2.3.3.3). It is possible that these latest developments will result in high-irradiance semiconductor lasers emitting at 630 nm, which will be very useful for photodynamic therapy.

3.7.2.4 Photodynamic Therapy Using HPD (HPD–PDT)

The efficiency of the HPD–PDT process will depend on several factors: (i) HPD concentration in the tumor, (ii) the light fluence inside the tumor, (iii) the extinction coefficient at the given wavelength, (iv) the quantum yield of the generation of a triplet state, (v) the presence of oxygen, and (vi) the quantum yield of singlet oxygen generation.

Some animal studies indicate that the blood flow in the tumor is decreased after the light treatment. In addition to the cytotoxic effect of singlet oxygen, HPD–PDT may cause microvascular shutdown that causes tumor death.

Clinical studies of HPD–PDT on more than 5000 patients have been performed worldwide in many centers over the last decade. The results of these studies are encouraging, and more details of these studies are given in Chapter 9.

HPD has several disadvantages. One is that patients develop unwanted skin reactions when exposed to sunlight. Other potential sensitizers have been proposed, including fluorescein, eosine, tetracycline, acidic orange, and various porphyrins. Many of these are currently under study as potential replacements for HPD.

3.7.2.5 Selective Excitation of Drugs—Photochemotherapy

The HPD–PDT exemplifies a whole family of medical laser applications. The method is based on the administration of a drug that is selectively retained in an organ. The drug must be triggered by optical radiation. This mechanism is often specific and the drug will be triggered only by a certain wavelength. This may be aptly called photochemotherapy. Although this method does not necessarily involve lasers, the triggering light must possess high irradiance and lasers are useful for this treatment.

3.7.3 Laser Therapy Due to Photomechanical Mechanisms

In Section 3.4.5 we discussed two types of photomechanical phenomena: (i) laser-induced breakdown in transparent media and the generation of plasma and shock waves and (ii) the generation of stress waves due to thermoelastic and ablative effects. Both phenomena have been used clinically in surgery and therapy. Some important examples are discussed.

3.7.3.1 Plasma and Shock Waves in Ophthalmology

Pulses generated by Nd:YAG or dye lasers have been focused inside the eye, with pulse duration <1 μsec and energies of tens of millijoules per pulse. Plasma, generated at the focal spot, produces a shock wave that destroys tissue in a selected site. The plasma produced at the target absorbs the Nd:YAG light and prevents further propagation of the laser beam, thus protecting the retina from damage.

This method has been used for several applications: (i) The destruction of opaque strands inside the eye. (ii) Iridotomy—generation of a hole in the iris in order to reduce excess pressure in the eye in glaucoma. (iii) Capsulotomy—after cataract removal and artificial lens implantation, an opaque capsule sometimes remains but can be destroyed by a focused laser beam without damage to the cornea or the retina.

3.7.3.2 Plasma and Shock Waves in Urology

Short, energetic pulses may be focused on the surface of kidney stones that are immersed in water. In a process similar to the one described earlier, plasma and shock waves are generated.

These phenomena were discovered in the late 1960s. In the mid-1980s, it was discovered that the energetic laser pulses could be delivered through thin optical fibers. This led to clinical fragmentation of stones, or laser lithotripsy (see Section 9.12.3). These laser pulses generate enormous power densities and high electric fields near the surface of the stone. These induce ionization of atoms and molecules and produce free electrons in a rapid avalanche process, leading to plasma formation. The temperature may rise to more than 10,000°C, accompanied by a bright flash of light. The plasma expands and induces a shock wave which is confined on one side by the liquid. It propagates through the stone, leading to mechanical disruption.

The Q-switched Nd:YAG laser induces optical breakdown in water, which is then coupled into the stone. Because the required high peak power is not easily transmitted by fibers, the process is less efficient. The pulsed dye laser breakdown is induced at the stone surface, leading to more efficient coupling. For dye laser energy transmitted through a 0.2-mm fiber, these phenomena occur at roughly 70 MW/cm². This energy density can be transmitted by fused silica fibers, which

FIGURE 3.13 Laser lithotripsy demonstration: removal of select areas of an egg shell. (Courtesy of Candela.)

are damaged only at power densities of 10^9–10^{10} W/cm^2, and this explains why pulsed dye laser lithotripsy works so well.

As an illustration of this phenomenon, such laser pulses were used to remove selective areas of the hard shell of an egg. The rest of the egg remained intact, as shown in Fig. 3.13.

3.7.3.3 Stress Waves in Corneal and Vascular Ablation

Stress waves are generated even when the power density does not generate plasma and shock waves (see Section 3.4.5.2). Such waves are generated when tissue is exposed, for example, to strongly absorbed laser pulses. They may be responsible, in part, for the ability of excimer laser pulses to remove corneal tissue with exquisite precision in corneal reshaping procedures. They may also play an important role in the ablation of vascular tissue during laser angioplasty.

Some of the information that was given in Sections 3.4 and 3.7 is now summarized in two tables. Table 3.2 shows the important medical lasers and some of

TABLE 3.2   **Medical Lasers and Their Applications**

Laser (general)	Wave-length (nm)	Laser (type)	Tissue effects	Discipline	Typical application
Excimer	193,249 308,351	Very short pulses	Ablation with little thermal damage	Ophthalmology Cardiology	Corneal surgery Laser angioplasty
Ar ion	488,514	CW	Coagulation Vaporization of lesions	Dermatology	Port wine stains Tattoo excision
				Gastroenterology Gynecology Otolaryngology	Bleeding lesions Menorrhagia Bleeding lesion
		Pulsed		Ophthalmology	Retinal reattachment Iridectomy Glaucoma
Dye	400–1000	CW	Sensitizer triggering	Oncology (+ others)	Photodynamic therapy
		Pulsed	Selective absorption	Dermatology	Port wine stains
			Plasma	Urology	Laser lithotripsy
Nd:YAG	1060	CW	Volume heating	Gastroenterology	Hemostasis Tumor cure
				General surgery	Cholecystectomy
		Pulsed	Plasma and shock	Urology Ophthalmology	Laser lithotripsy Tumor destruction
CO_2	10.6 μm	CW, long pulses	Precise cutting	Surgery	Tissue removal Laser surgery

TABLE 3.3 Medical Applications of Lasers

Field	Power density	Dura- tion	Depth of pene- tration	Medical applications	Example
Diagnosis	Very low	Long	Shallow	Blood diagnosis Tissue characterization	HeCd
Therapy	Low	Long	Deep	Biostimulation	HeNe
	Medium	Long	Medium	Tissue welding Blood coagulation	Nd:YAG Ar ion
	Medium	Long	Deep	Laser hyperthermia Phototherapy	Nd:YAG Au vapor
	High	Short		Laser lithotripsy	Dye
Surgery	High	Long	Shallow	Cutting, ablation	CO_2
	Very high	Very short	Shallow	Ablation without thermal damage	Excimer Er:YAG

the typical applications of each laser. Table 3.3 shows the important medical applications and some of the typical lasers that have been used for each application.

3.8 THERMAL INTERACTION BETWEEN LASER AND TISSUE—ADVANCES

Thermal interaction between lasers and tissue has been the subject of extensive experimental and theoretical work, described in detail elsewhere (Welch, 1984; Welch *et al.*, 1991; McKenzie, 1990; Jacques, 1990, 1992). Experimentally, surgeons used at first CO_2, Nd:YAG, or Ar lasers, which were operated continuously (CW) or in long pulses. Microscopic observation and histopathology of the laser cut showed an extensive region of thermal damage. Later it was found that cuts made by excimer or Er:YAG lasers showed minimal thermal damage. Theoretically, scientists have used sophisticated calculations that explain the difference between the various types of interactions between lasers and tissue. These theoretical methods are illustrated in this section by a somewhat simplified theoretical calculation.

3.8.1 Temperature Rise in Tissue

When a sample is heated by a heat source, its temperature T rises and it is possible to calculate the spatial and temporal changes of T. Let us consider the case in which heat is generated at the upper surface of the sample and there are no losses due to convection or radiation. The behavior of the sample is governed (McKenzie, 1990) by the general heat conduction equation:

$$\partial T(x,\ y,\ z,\ t)/\partial t\ =\ K\ \nabla^2 T(x,\ y,\ z,\ t)\ +\ Q(x,\ y,\ z,\ t)/\rho c. \qquad (3.9)$$

In the one-dimensional case, this equation has the form

$$\partial T(z, t)/\partial t = K\, \partial^2 T(z, t)/\partial z^2 + Q(z, t)/\rho c, \qquad (3.10)$$

where $K = k/\rho c$ ($cm^2\ sec^{-1}$) in the thermal diffusivity, k is the thermal conductivity ($cal\ cm^{-1}\ sec^{-1}\ deg^{-1}$), c is the specific heat ($cal\ gm^{-1}\ deg^{-1}$), and ρ is the density ($g\ cm^{-3}$). Q is the heat source term ($cal\ cm^{-3}\ sec^{-1}$) and it specifies the rate of heat generated in a unit volume of the sample.

Consider a flat sample whose surface is the x, y plane. A Gaussian laser beam of the type given in Eq. (2.1) is directed along the z axis, perpendicular to the sample surface. We may replace Eq. (3.2) by the following equation:

$$I(r, z) = I_0(r)\, \exp(-\alpha z). \qquad (3.11)$$

If the penetration depth $1/\alpha$ is much smaller than the beam diameter on the surface, we can ignore the dependence on x and y and use Eq. (3.10). In this case the absorbed energy generates heat and the rate of heat generated per unit volume is given by

$$Q = \partial I/\partial z = \alpha I. \qquad (3.12)$$

When this is substituted in Eq. (3.9), it may in principle be solved numerically for each of a number of totally different irradiation conditions:

(i) Laser wavelengths for which there is deep penetration into tissue and strong scattering (e.g., Nd: YAG or GaAs).

(ii) Laser wavelengths that are strongly absorbed with practically no scattering (e.g., excimer, Er: YAG, or CO_2).

(iii) Long pulses (or CW), where heat dissipation occurs via conduction during the pulse.

(iv) Very short pulses, of duration t_s, where there is practically no dissipation during the pulse.

It is difficult to calculate $T(z, t)$ in some of these cases. The most easily solved case involves a very short pulse of highly absorbed laser radiation. All the laser energy is absorbed in the surface layer. The temperature rises rapidly to a final value T_m that is proportional to the deposited energy. There is no heat loss during the pulse. At the end of the pulse the temperature decreases with a time constant τ_r, called the relaxation time. This is often written as follows (Welch et al., 1991):

$$\tau_r = L^2/4K, \qquad (3.13a)$$

where L is a characteristic length. In the case of strong absorption one substitutes $L = 2/\alpha$, where α is the absorption coefficient [see Eq. (3.2)] and therefore

$$\tau_r = (1/\alpha)^2/K. \qquad (3.13b)$$

EXAMPLE 1—*Single Pulse*: For tissue K is of the order of $10^{-3}\ cm^2/sec$, and therefore when tissue is exposed to a CO_2 laser beam, L is roughly $2 \times 10^{-3}\ cm$

and τ_r is roughly 1 msec. Therefore in (iv) above, the duration t_s of the short laser pulse must be $t_s < 1$ msec to ensure negligible thermal diffusion during the laser pulse and negligible accumulation of heat energy in the surrounding tissue.

EXAMPLE 2—*Train of Pulses*: Let us assume that the tissue is exposed to a train of pulses at a repetition rate f pulses per second. The time interval between two consecutive pulses is $1/f$, and thus if $1/f > \tau_r$ the tissue has sufficient time to cool down between the two pulses. In Example 1 the repetition rate f has to be much less than 1000 Hz (e.g., $f = 100$ Hz). There are waveguide CO_2 lasers that operate at much higher repetition rates (e.g., 10,000 Hz). In such cases the tissue temperature increases gradually, and the situation is as if the tissue was exposed to a CW laser beam.

3.8.2 Thermal Damage in Tissue

Exposure of tissue to a high temperature for a relatively long period of time causes irreversible damage. The damage may be defined as denaturation of protein or loss of function. It can be shown that the damage may be described mathematically by a damage function Ω defined as

$$d\Omega/dt = A \, \exp(-E/RT). \tag{3.14}$$

In this equation, A and E are constants and R is the universal gas constant. Typically, complete tissue necrosis occurs if $\Omega \geq 1$ and no damage occurs if $\Omega < 0.5$. The constants E and A have been estimated by Henriques and Moritz from experiments on pig skin (Welch, 1985).

In practice, one "sets" $\Omega = 0.5$ as the damage threshold. From experimental data, one calculates E and A values that yield $\Omega > 0.5$ for every T–t combination that shows damage experimentally.

EXAMPLE: If tissue temperature rises rapidly to a value T and stays for a period t_p at that temperature (e.g., by application of a laser pulse), one could write

$$\Omega = At_p \, \exp(-E/RT). \tag{3.15}$$

Let us assume now that the tissue's initial temperature is $T_i = 37°C$. For each temperature rise ΔT (and final temperature $T = T_i + \Delta T$), one could calculate the exposure time t_p for which thermal damage occurs. The results of such calculations for several temperatures are shown in Figure 3.14. It is clear that the thermal damage is sensitive to the final tissue temperature. For example for $\Delta T = 14°C$ one obtains $t_p = 100$ sec, whereas for $\Delta T = 20°C$ the corresponding value is $t_p = 1$ sec. This figure also explains why at a temperature of 60°C coagulation necrosis occurs in roughly 1 sec (see Section 3.7.1.4).

3.8.3 Ablation Parameters of Tissue

The foregoing calculation is valid as long as the pulse duration is short enough that thermal diffusion can be ignored, as discussed in Section 3.8.1. The

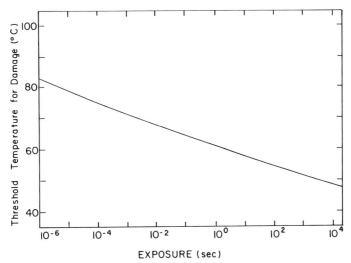

FIGURE 3.14 Tissue temperature T and time t_p to reach thermal damage at this temperature.

same calculation can be performed for the three types of lasers: excimer, Er:YAG, and CO, all of which have high absorbance in water. In place of water, similar results may be constructed for real tissue. For example, the theoretical results for an atheroma are shown in Table 3.4.

Comment: For each wavelength λ there is a corresponding absorption coefficient α. 63% of the laser energy is absorbed within a distance $1/\alpha$ and 90% of the energy within a distance of $2.3/\alpha$. In Table 3.4 we followed previous authors (Bonner *et al.*, 1986) and assumed that $L = 2/\alpha$ and obtained, for example for a CO_2 laser, a threshold (minimum) fluence $F_{th} = 5$ J/cm² for ablation. On the other

TABLE 3.4 **The Physical Parameters of Atheroma**

Laser	Wavelength λ	Absorption coefficient α (cm⁻¹)	Penetration depth $1/\alpha$ (μm)	Fluence threshold (J/cm²)	Relaxation time (msec)
KrF	248 nm	5–600	15–20	5	1
XeCl	308 nm	200–250	40–50	10	5
Dye	400–500 nm	30–50	200–350	50–70	70–80
Ar	514 nm	30	350	70	80
Nd:YAG	1.06 μm		Millimeters		
Ho:YAG	2.12 μm	25–35	300–400	70–100	80
Er:YAG	2.94 μm	3–5000	2–3	0.5	0.08
CO_2	10.6 μm	600–1000	15–20	5–10	1

hand, if one assumes $L = 4/\alpha$, then for CO_2 the value $F_{th} = 10$ J/cm^2 is obtained. This is the value that is used in Table 3.1

Train of pulses: Let us consider a pulsed CO_2 laser. Each consecutive pulse, in a train of pulses, will vaporize a thickness of 0.005 cm. If the repetition rate is 100 Hz (100 pulses/sec), the laser pulses vaporize $100 \times 0.005 = 0.5$ cm/sec. The average "thickness" removal rate (or cutting rate) is thus 5 mm/sec.

3.8.4 Optimum Conditions for Laser Surgery

This section considers a simplified model of a laser beam that is absorbed in tissue and whose energy is converted to heat. When the temperature of the heated area is above the vaporization point, the tissue is vaporized. Neglected here are chemical bond breaking, plasma formation, and acoustic effects. We also assume that, for the laser absorption, one could use the absorbance and the penetration depths that are normally measured at low power levels.

During the laser pulse, the energy generated may be diffused to neighboring areas. The characteristic time constant for the energy loss is the relaxation time given in Eq. (3.13). If the pulse length is shorter than τ_r, there is practically no loss and all the energy can be used for tissue vaporization. For the CO_2 laser, this will happen for pulse lengths $t_p < 0.25$ msec.

When the wavelength chosen is in the visible, the energy is deposited deep in tissue, where scattering effects are noticeable. On the other hand, for wavelengths in the UV or mid-IR, the energy is deposited in a thin surface layer, and therefore the deposited energy will be efficiently utilized for vaporization. For the CO_2 laser, for example, the minimum energy needed for ablation (i.e., ablation threshold) is only 10 J/cm^2, whereas for Nd:YAG it may be 10 times higher.

For all pulse lengths shorter than τ_r, we expect the same results. On the other hand, the pulse energy must be above the ablation threshold energy density. If one uses short pulses, the peak pulse power tends to increase. Short pulses of high peak power cannot be easily transmitted through thin optical fibers (see Section 8.4). Therefore pulses much shorter than τ_r should not be used.

Figure 3.15 illustrates the theoretical results obtained for the interaction between two types of laser beams and tissue: (i) long pulsed Gaussian beam and (ii) short pulsed "top hat" beam. In both cases we see three zones in the tissue: a zone where the temperature is higher than the ablation temperature and tissue is removed, a zone where the temperature is lower than the ablation temperature but thermal damage is caused, a zone where heating occurs without causing thermal damage. The main difference between the two cases is the extent of the thermal damage zone.

Figure 3.16a shows microphotographs of tissue that was exposed to three different lasers: YAG, CO_2, and excimer. Figure 3.16b shows the histological cross sections of tissue that was exposed to several types of lasers: Ar, Nd:YAG, and excimer. The long duration Gaussian beams cause thermal damage, whereas the short pulsed excimer laser causes little damage.

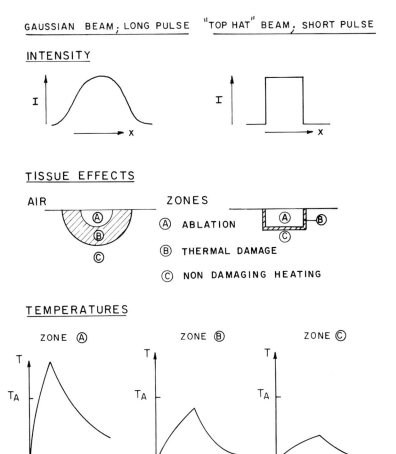

FIGURE 3.15 Tissue effects of Gaussian beam versus "top hat" beam—theoretical model.

We may therefore conclude that the optimal laser parameters for laser surgery without thermal damage are

(i) Wavelength that is well absorbed in tissue
(ii) Pulse length roughly equal to τ_r
(iii) Pulse energy density above the ablation threshold
(iv) A top hat irradiance distribution (as shown in Fig. 3.15), such as the excimer laser, for which thermal damage is lower than for a Gaussian distribution
(v) A repetition rate sufficiently low to allow tissue cooling between consecutive pulses

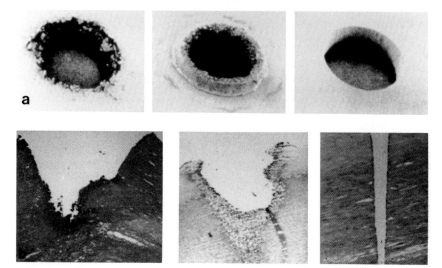

FIGURE 3.16 (a) Microphotograph of tissue effects of various laser beams. (Courtesy of Advanced Interventional Systems.) (b) Histology of tissue effects of various laser beams. (Courtesy of Advanced Interventional Systems.)

References

Alfano, R. R., and Doukas, A. G. (1984). Special issue on Lasers in Biology and Medicine. *IEEE J. Quantum Electron.* **QE-20**, No. 12.

Alfano, R. R., Lam, W., Zarrabi, M. A., Alfano, M. A., Cordero, J., Tata, D., and Swenberg, C. (1984). Human teeth with and without caries studied by laser scattering, fluorescence and absorption. *IEEE J. Quantum Electron.* **QE-20**, 1512–1516.

Andersson-Engels, S., Johansson, J., Svanberg, S., and Svanberg, K. (1989). Fluorescence diagnosis and photochemical treatment of diseased tissue using lasers: part I. *Anal. Chem.* **61**, 1367a–1373a.

Andersson-Engels, S., Johansson, J., Svanberg, S., and Svanberg, K. (1990). Fluorescence diagnosis and photochemical treatment of diseased tissue using lasers: part II. *Anal. Chem.* **62**, 19a–27a.

Atsumi, K. (1983). *New Frontiers in Laser Medicine and Surgery.* Amsterdam: Excerpta Medica.

Bass, L. S., Moozami, N., Pocsidio, J., Oz, M. C., Logerfo, P., and Treat, M. R. (1992). Changes in type I collagen following laser welding. *Laser Surg. Med.* **12**, 500–505.

Berns, M. W. (1988). *Laser Interaction with Tissue,* SPIE Proceedings, Vol. 908. Bellingham, WA: SPIE.

Berns, M. W. (1991). Laser surgery. *Sci. Am.* **264**, 84–90.

Berry, M. J., and Harpole, G. M. (1989). *Thermal and Optical Interactions with Biological and Related Composite Materials,* SPIE Proceedings, Vol. 1064. Bellingham, WA: SPIE.

Birngruber, R. (1990). Special issue on lasers in biology and medicine. *IEEE J. Quantum Electron.* **QE-26**, No. 12.

Bonner, R. F., Smith, P. D., Leon, M. B., Esterowitz, L., Storm, M., Levin, K., and Tran, D. (1986). Quantification of tissue effects due to a pulsed Er: YAG laser at 2.9 μm with beam delivery in a wet field via zirconium fluoride fibers. *Proc. SPIE* **713**, 2–5.

Boulnois, J. L. (1986). Photophysical processes in recent medical laser developments: a review. *Lasers Med. Sci.* **1**, 47–66.

Bown, S. G. (1986). Laser endoscopy. *Br. Med. Bull.* **42**, 307–313.

Caro, R. C., and Choy, D. S. J. (1992). Special issue: optics and light in medicine. *Opt. Photon. News* **3** (10), 11–44.

Carruth, J. A. S., and McKenzie, A. L. (1986). *Medical Lasers: Science and Clinical Practice.* Bristol, England: Adam Hilger.

Council on Scientific Affairs. (1986). Lasers in medicine and surgery. *JAMA* **256**, 900–907.

Deutsch, T. F. (1988). Medical applications of lasers. *Phys. Today* **41** (10), 56–63.

Deutsch, T. F., and Puliafito, C. A. (1987). Special issue on lasers in biology and medicine. *IEEE J. Quantum Electron.* **QE-23**, No. 10.

Deutsch, T. F., and Wynant, R. (1992). Special section on lasers in medicine. *Proc. IEEE* **80** No. 6.

Dixon, J. A. (1987). *Surgical Applications of Lasers*, 2nd ed. Chicago: Year Mook Medical Publishers.

Dougherty, T. J. (1989). Photodynamic therapy: status and potential. *Oncology* **3**, 67–73.

Edelson, R. L. (1988). Light activated drugs. *Sci. Am.* **259**, 50–57.

Goldman, L. (1991). *Laser Non Surgical Medicine.* Lancaster: Tachnomic.

Hebden, J. C., Kruger, R. A., and Wong, K. S. (1991). Time resolved imaging through a highly scattering medium. *Appl. Opt.* **30**, 788–794.

Hecht, E. (1987). *Optics*, 2nd ed. Reading, MA: Addison-Wesley.

Henderson, B. W., and Dougherty, T. J. (1992). *Photodynamic Therapy.* New York: Marcel Dekker.

Jacques, S. L. (1990). *Laser–Tissue Interaction*, SPIE Proceedings, Vol. 1202. Bellingham, WA: SPIE.

Jacques, S. L. (1991). *Laser–Tissue Interaction II*, SPIE Proceedings, Vol. 1427. Bellingham, WA: SPIE.

Jacques, S. L. (1992). *Laser–Tissue Interaction III*, SPIE Proceedings, Vol. 1646. Bellingham, WA: SPIE.

Joffe, S. N., and Atsumi, K. (1990). *Laser Surgery: Advanced Characterization, Therapeutics and Systems II*, SPIE Proceedings, Vol. 1200. Bellingham, WA: SPIE.

Joffe, S. N., Goldblatt, N. R., and Atsumi, K. (1989). *Laser Surgery: Advanced Characterization, Therapeutics and Systems*, SPIE Proceedings, Vol. 1066. Bellingham, WA: SPIE.

Kessel, D. (1990). *Photodynamic Therapy of Neoplastic Diseases.* Boca Raton, FL: CRC Press.

Kittrell, C., Willett, R. L., de los Santos-Pancheo, C., Ratliff, N. B., Kramer, J. R., Malk, E. G., and Feld, M. S. (1985). Diagnosis of fibrous arterial atherosclerosis using fluorescence. *Appl. Opt.* **24**, 2280–2281.

Leon, M. B., Lu, D. Y., Prevosti, L. G., Macy, W. W., Smith, P. D., Granovsky, M., Bonner, R. F., and Balaban, R. S. (1988). Human arterial surface fluorescence: atherosclerotic plaque identification and effects of laser atheroma ablation. *J. Am. Coll. Cardiol.* **12**, 94–102.

Levi, L. (1968). *Applied Optics*, Vol. 1, New York: Wiley.

Marcus, S. L. (1992). Photodynamic therapy of human cancer. *Proc. IEEE* **80**, 869–889.

Martellucci, S., and Chester, A. N. (1989). *Laser Photobiology and Photomedicine.* Ettore Majorana Int. Sci. Ser.

McGuff, P. E., Bushnell, D., Sorroff, H. F., and Deterling, R. A. (1963). Studies of the surgical applications of laser. *Surg. Forum* **14**, 143–145.

McKenzie, A. L. (1990). Physics of thermal processes in laser–tissue interaction. *Phys. Med. Biol.* **35**, 1175–1209.

Oster, G. (1986). Laser surgery. In: Garetz, B. A., and Lombardi, J. R. (Eds.), *Advances in Laser Spectroscopy*, Vol. 3. New York: Wiley.

Parrish, J., and Deutsch, T. F. (1984). Laser photomedicine. *IEEE J. Quantum Electron.* **QE-20**, 1386–1396.

Parrish, J. A., and Wilson, B. C. (1991). Current and future trends in laser medicine. *Photochem. Photobiol.* **53**, 731–738.

Pratesi, R. (1991). *Optronic Techniques in Diagnostic and Therapeutic Medicine.* New York: Plenum Publishing.

Preuss, L. E., and Profio, A. E. (1990). Special issue: optical properties of mammalian tissue. *Appl. Opt.* **28**, 2207–2357.

Prince, M. R., Anderson, R. R., Deutsch, T. F., and LaMuraglia, G. M. (1988). Pulsed laser ablation of calcified plaque. *Proc. SPIE* **906**, 305–309.

Regan, J. D., and Parrish, J. A. (1982). *The Science of Photomedicine*. New York: Plenum Publishing.

Spears, J. R., Serur, J., Shropshire, D., and Paulin, S. (1983). Fluorescence of experimental atheromatous plaques with hematoporpyrin derivative. *J. Clin. Invest.* **71**, 395–399.

Steger, A. C., Lees, W. R., Walmsley, K., and Bown, S. G. (1989). Interstitial laser hyperthermia: a new approach to local destruction of tumours. *Br. Med. J.* **299**, 362–365.

Thomsen, S. (1991). Pathologic analysis of photothermal and photomechanical effects of laser–tissue interactions. *Photochem. Photobiol.* **53**, 825–835.

van Gemert, M. J. C., and Welch, A. J. (1989). Clinical use of laser–tissue interactions. *IEEE Eng. Med. Biol.* **8** (4), 10–13.

Welch, A. J. (1984). The thermal response of laser irradiated tissue. *IEEE J. Quantum Electron.* **QE-20**, 1471–1481.

Welch, A. J. (1985). Laser irradiation of tissue. In: Shitzer, A., and Everhart, R. C. (Eds.), *Heat Transfer in Medicine and Biology*, Vol. II. New York: Plenum Publishing.

Welch, A. J. (1989). Special issue on laser tissue interaction. *IEEE Trans. Biomed. Eng.* **36**, 1145–1242.

Welch, A. J., Motamedi, M., Rastegar, S., LeCarpentier, G. L., and Jansen, D. (1991). Laser thermal ablation. *Photochem. Photobiol.* **53**, 815–823.

Wilson, B. C, Sevick, E. M., Patterson, M. S., and Chance, B. (1992). Time dependent optical spectroscopy and imaging for biomedical applications. *Proc. IEEE* **80**, 918–930.

Wilson, B. C. (1991). Lasers in medicine. *Photochem. Photobiol.* (special issue) **53**.

Yahr, W. Z., and Strully, K. J (1966). Blood vessel anastomosis by laser and other biomedical applications. *J. Assoc. Adv. Med. Instrumen.* **1**, 28.

4

Single Optical Fibers

4.1 INTRODUCTION

Chapters 2 and 3 outlined the principles of laser science, its potential medical applications, and the various lasers that have been used in medicine. These laser applications could be used to fuller advantage if it were possible to deliver the laser beam inside the body. Two decades ago, with the help of a set of mirrors—known as an articulated arm—many laser treatments were implemented. These mirror systems, however, were bulky and cumbersome to use. The optical device that enables physicians to perform laser treatment more efficiently is the optical fiber. Sufficiently small-diameter optical fibers act as guides to transmit the laser beam inside the body, for both diagnostic and therapeutic applications. Arrays of fibers—fiberoptic bundles—are the basic components of endoscopes that are widely used for imaging inside the body. The subject of this chapter is that of single optical fibers.

4.2 HISTORICAL BACKGROUND

Ancient glassblowers in Greece and elsewhere were well aware of the transmission of light through thin rods or threads of glass. This phenomenon was probably used for special lighting effects in decorative glassware. A clear demonstration of this effect was given in the 19th century by the famous British scientist of Irish descent, John Tyndall. Like others in academia, Tyndall experienced difficulty obtaining a tenured position in a respectable university. Thus he earned his living by giving lectures, accompanied by illustrative experiments, all over Britain. To illustrate one of his talks, he prepared an experiment using a transparent container filled with water. The container had a small hole on the side wall near the bottom. When light was transmitted through the container into the hole, it was "trapped" inside the flowing stream of water by a series of internal reflections. John Tyndall, in a lecture to the Royal Institution in London in 1854, "permitted water to spout from a tube, the light on reaching the limiting surface of air and water was totally reflected and seemed to be washed downwards by the descending liquid" (Tyndall, 1854, as cited in Allan, 1973). The experiment is illustrated schematically in Fig. 4.1a. Tyndall's experiment may easily be demonstrated today with the help of an Ar ion or HeNe laser and a stream of water. An illustration is given in Fig. 4.1b.

The same principle of total internal reflection applies to the transmission of light in thin threads of glass. The technology for making such threads was available in England at that time. Charles Boys (of the soap bubbles fame) was a distinguished physician who tried to fabricate thin glass tubes for use as mercury thermometers. In 1887, he devised an interesting way to make these tubes. He attached a glass tube to an arrow which was mounted on a crossbow. He then heated the tube near its end until it reached its melting point and released the arrow. The arrow pulled the glass, which formed a thin hollow tube. An almost identical technique is used today for fabricating glass fibers and even the velocities involved are similar. Nevertheless, scientists and engineers of the period were apparently not interested in transmitting light through a gossamer thread of glass and the field lay dormant for many decades.

Modern optical fiber technology began in 1954 in England, when Prof. H. H. Hopkins, with his student N. S. Kapany, used fiber bundles for imaging and started a series of studies on the properties of single optical fibers (Hopkins and Kapany, 1954). The major disadvantage of these imaging devices was that the individual fibers made in England were uncoated. Light "leaked" from fiber to fiber, and thus they were not suitable for producing good imaging devices. Simultaneously, in Holland, Abraham van Heel realized the importance of fibers consisting of an inner part or core and an outer cladding layer made of a different material (van Heel, 1954). In his early work, the core was made of glass, the cladding layer was made of plastic, and the optical quality was rather poor. Although such fibers were unsuitable for practical applications, they did demonstrate the importance of cladding layers for single fibers.

The major breakthrough was made in 1956 in the United States by Curtiss, Hirschowitz, and Peters. They discovered a way to make clad fibers by inserting a rod of glass into a tube of a different type of glass. The end of this compound rod was then heated and pulled. The procedure thus formed a thin fiber with glass core and glass cladding (Curtiss et al., 1957). Although this paved the way for both the fabrication of ordered imaging bundles and the development of endoscopes, as described in Chapter 6, the optical quality of the fibers was still poor (Kapany, 1960, 1967). They were sufficient for transmitting light over short distances, as required for endoscopy, but the optical losses were high.

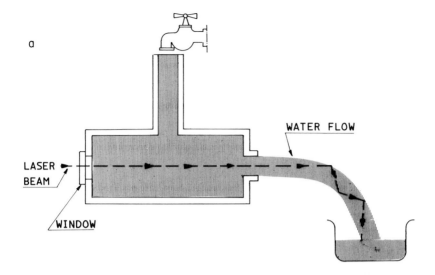

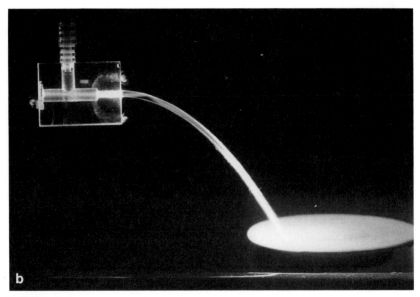

FIGURE 4.1 Tyndall experiment: light transmission through water. (a) Schematic drawing; (b) photograph.

The next phase in the development of optical fibers is linked to optical communications. In the United States, Alexander Graham Bell, the inventor of the telephone, understood the possibility of sending messages via light beams (Hecht, 1985). In 1880, he invented the photophone, a crude system that made it possible to send voice communication via a beam of sunlight. The beam was reflected off a thin membrane to a light detector that was placed at a distance. The human voice made the membrane vibrate, thus changing the intensity of light reaching the detector. These changes were converted to electrical signals, which were used to drive a loudspeaker to reproduce the voice. Using this system, Bell was able to transmit a voice from one hill to another, through the atmosphere, covering a distance of about 1 mile. Its main problem, however, was that it was inoperable in foggy or dusty atmospheres. The idea of transmitting light beams through transparent fibers for communication was not tried, even though Bell probably knew of Tyndall's experiments. In any event, Bell considered the invention of the photophone as even more significant than that of the telephone—and this may well turn out to be true.

With the development of the first laser in 1960, people started to reconsider the notion of optical communication. They conceptualized an idea which involved sending a laser beam through a fiber and modulating (i.e., changing) its intensity with a human voice. The light intensity would go up and down in a coded way—corresponding to the voice. A detector placed at the end of the fiber would pick up the modulated light and send out corresponding electrical signals. These signals would carry similar information in identical codes, and if these signals were connected to a suitable electronic system and a loudspeaker, the voice would be regenerated. As mentioned, however, the optical quality of the fibers left much to be desired. It was pointed out (Kao and Hockham, 1966) that purifying glass could dramatically improve its transmission properties. Indeed, using techniques developed for silicon purification, the transmission of glass improved within a few years. Modern glass fibers are highly transparent. More than 95% of the light incident on a fiber whose length is 1 km is transmitted and only 5% is scattered or absorbed (neglecting end-face reflection).

These fibers are the building blocks of communication systems, some of which are being tested on a wide scale. A miniature semiconductor laser, whose size is a fraction of a millimeter, will be coupled to one end of such a fiber. The laser light will be modulated by voice, video, or computer signals. The light transmitted through the fiber will be detected at the other end by a miniature semiconductor detector. The signal from this detector will then be transformed again to a voice, TV, or computer signal. The major advantage of optical fibers is their ability to transmit enormous amounts of information with minimum loss. Currently, tens of TV programs or thousands of telephone calls can be delivered simultaneously through one fiber! There is little doubt that optical fibers will replace copper wires in all future communication systems (Allan, 1973; Cherin, 1983).

This revolution in fiberoptic communication is having a direct effect on medicine. Present-day copper wire systems are not suitable for the transmission of substantial amounts of information. Optical fibers will pave the way for telemedicine. One application is communication; physicians in a particular location will examine a patient and record the results by a video system. They will also perform some sophisticated diagnosis, using x-rays, ultrasound, or magnetic resonance imaging (MRI). The video image and the diagnosis results will be transmitted in real time to a central location, where experts may be consulted. A second application is in information storage and retrieval. Computers cannot "talk" efficiently to each other via copper wires. In the future, computers used by hospitals will be able to communicate with each other or with large computers in central locations (such as central files and large libraries) via optical fibers. Medical information, such as the medical history of patients or the diagnosis of a certain disease, will thus be readily available. While sitting at their office computers, physicians will be able to retrieve this medical information or scientific information from central libraries. Today, written material, graphs, ordinary or x-ray pictures, MRI records, or computed tomographic scan information are not easily accessible and may be lost. In the future, this information will be stored in digital form (e.g., on compact discs) and available for instantaneous retrieval via the optical fiber network.

The rapid developments in optical communications have generated worldwide interest in the manufacture of high-quality optical fibers. Research and development are being conducted in both research institutions and the large industrial laboratories. Concurrently, other applications of optical fibers have also been rapidly developed, including various types of diagnostic procedures. Optical fibers can also transmit high-power laser beams for laser therapy inside the body. Special optical fibers have been developed for transmitting optical radiation either in the ultraviolet (UV) or in the middle and far infrared (IR). This progress has had a great impact on the field of fiberoptics in medicine. First and foremost was the use of better fibers for endoscopy, which resulted in higher-resolution images with near-perfect color rendition and improved mechanical properties. Simultaneously, new fiberoptic techniques paved the way for using optical fibers in systems such as laser catheters or laser endoscopes (Katzir, 1991 and 1992).

The development of optical fibers is a culmination of many decades of research and development that was conducted by a plethora of scientists and engineers. Yet it must be mentioned at this point that nature was already using this phenomenon millions of years ago. Biologists discovered that some tissues of plant seedlings act like optical fibers that guide light, helping the plants to coordinate their physiology. In the animal world, there are optical guiding phenomena in imaging systems such as the eyes of many animals, including the human eye. Finally there are reports that UV-transmitting or IR-transmitting optical fibers also appear in nature (see Section 4.7).

4.3 OPTICAL FIBERS—FUNDAMENTALS

When light travels from air into water or glass, a small fraction of the light is reflected back into the air, while the rest is transmitted into the water or glass. These phenomena are called external reflection and refraction. The situation is rather similar if light that is transmitted from glass (or water) into the air is nearly perpendicular to the interface between the glass and the air. However, if a light beam is sent from glass to air at an angle of about 45° or more with respect to the

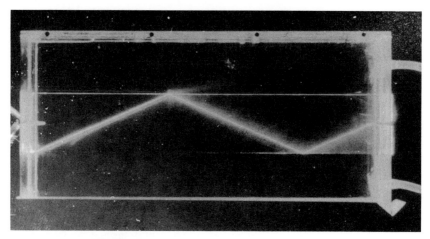

FIGURE 4.2   Photograph of light propagation in a glass slab.

normal to the interface, a new phenomenon occurs. The light is totally reflected back into the glass. This is called total internal reflection (Hecht, 1987).

This can be beautifully demonstrated by using a thick slab of glass (or Lucite) and sending a beam of light at an appropriate angle. The beam will reach the interface between the glass and the air and will be totally internally reflected. It will travel through the glass slab and, on reaching the other side, it will be reflected again. The beam will thus make a zigzag path inside the slab without ever escaping into the air. This is shown in Fig. 4.2, where a beam from an HeNe laser is depicted bouncing inside a slab of glass. It is clear that the beam travels from one side of the slab to the other side by a series of internal reflections. This is the same phenomenon as the transmission of light in a stream of water (as demonstrated by Tyndall). Incidentally, we see the beam only because some light is scattered toward our eyes.

The thick slab of glass may now be replaced by a thin, transparent, flexible rod of glass known as an optical fiber. Light can be transmitted through such a fiber by internal reflections, even when the fiber is bent or flexed. The phenomenon of total internal reflection of light at an interface is the basic principle underlying the ability of optical fibers to transmit light. These optical fibers are the building blocks of all laser–fiber systems (Seippel, 1984; Wolf, 1979).

4.4 LIGHT TRANSMISSION IN OPTICAL FIBERS—PRINCIPLES

The physical principles of light transmission through optical fibers have been reviewed in several books (Allan 1973; Allard, 1990). In this section we discuss few of these principles.

4.4.1 Total Internal Reflection

Although the transmission of light in optical fibers is a complex problem in physical optics, the phenomenon can be simply understood with an elementary model, the geometrical optics model. This model, which is often taught in high school, holds as long as the physical dimensions of the optical element are larger than the wavelength λ of light. In this book, almost all the dimensions of the various optical elements (including the fibers) are larger than λ, and the geometrical model is valid.

Consider two transparent media, medium 1 and medium 2, of refractive indices n_1 and n_2, respectively, with $n_1 > n_2$. A ray of light propagates at an angle Θ_1, with respect to the normal interface between the two media. The intensity[9] of this beam is represented by I_i. At its interface, the beam is divided into two. One part with an intensity I_r is reflected back into medium n_1, where the reflection may be specular for a smooth surface or diffuse for a rough surface. Another part, with an intensity I_t, is transmitted into medium n_2 and propagates at an angle Θ_2. The reflected and the refracted beams are shown separately in Fig. 4.3.

[9]The more accurate term here should be irradiance.

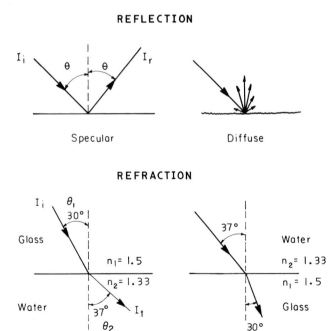

FIGURE 4.3 Reflection and refraction.

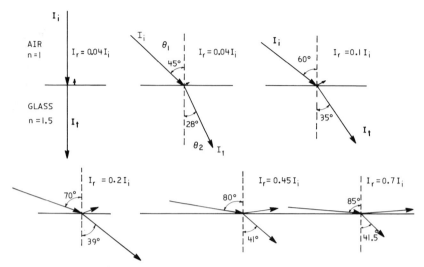

FIGURE 4.4 External reflection—glass–air interface.

There are three basic laws of geometrical optics (Hecht, 1987) for the case of specular reflection and transmission:

(i) The incident, reflected, and transmitted beams are all in one plane.

(ii) Reflection—the angle of incidence $\Theta_i = \Theta_1$ is equal to the angle of reflection Θ_r.

(iii) The refraction obeys Snell's law:

$$n_1 \sin \Theta_1 = n_2 \sin \Theta_2. \tag{4.1}$$

When the beam propagates in air and impinges on glass, we note that $n_1 < n_2$, and transmission and external reflection are observed for several values of Θ_1, as shown in Fig. 4.4. The amount of reflected light increases as Θ_1 increases. This is indicated by the lengths of the arrows representing I_r in Fig. 4.4.

If a beam propagates in glass and impinges on air, $n_1 > n_2$. Transmission and internal reflection are observed, as shown in Fig. 4.5.

According to Snell's law, if the incidence angle Θ_1 is increased, angle Θ_2 will also increase. Eventually $\sin \Theta_1$ may reach some value Θ_{1c} at which $\Theta_2 = 90°$. This angle Θ_{1c} is called the critical angle and is shown in Fig. 4.5. For this specific angle, Snell's law can be written as follows:

$$n_1 \sin \Theta_{1c} = n_2 \sin 90° = n_2. \tag{4.2}$$

For every angle of incidence which is larger than Θ_{1c}, there can be no refracted beam. To prove this, assume that there had been such a beam; its angle would have been given by the following equation:

$$\sin \Theta_2 = (n_1 \sin \Theta_1)/n_2 > (n_1 \sin \Theta_{1c})/n_2 = 1 \quad \text{or} \quad \sin \Theta_2 > 1.$$

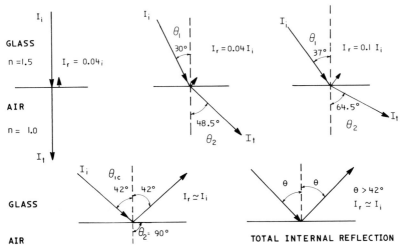

FIGURE 4.5 Internal reflection—glass–air interface.

Since this is physically impossible, it follows that only a reflected beam exists. This phenomenon is also shown in Fig. 4.5 and is called total internal reflection (TIR).

EXAMPLE: Consider the case in which medium 1 is glass with $n_1 = 1.5$ and medium 2 is air with $n_2 = 1$. The critical angle is

$$\sin \Theta_{1c} = 1/1.5 = 0.67 \quad \text{and} \quad \Theta_{1c} = 42°.$$

This total internal reflection is a common phenomenon in everyday life. A clear reflection in a regular window can easily be seen if the window is viewed from a shallow angle (larger than the critical angle). It also gives rise to reflections in prisms and is used in binoculars and other optical instruments.

EXAMPLE: If instead of air–glass the water–glass interface is considered, the same calculation can be repeated with $n_1 = 1.5$ and $n_2 = 1.33$. In this case the following is obtained:

$$\sin \Theta_{1c} = 1.33/1.5 = 0.86 \quad \text{and} \quad \Theta_{1c} = 60°.$$

For reflection of light from a surface, the reflectance R may be defined as the ratio

$$R = I_{\text{ref}}/I_{\text{in}}$$

between the intensities of the reflected beam and the incident one. Under conditions of total internal reflection ($\Theta > \Theta_{1c}$), the reflectance is extremely high and can easily reach $R = 0.9999995$. $I_i \approx I_r$ as indicated in Fig. 4.5. A real surface is not perfectly flat, and therefore R is slightly less than 1. This reflectance is much higher than the reflection from metals and is the main reason why optical fibers are so efficient in transmitting light.

4.4.2 Transmission of Light in a Straight Transparent Slab

Consider a long slab of transparent material that consists of three layers: an inner layer with a refractive index of n_1 and two outer layers with refractive indices n_2, with $n_2 < n_1$, as shown in Fig. 4.6. Several rays of light incident on the end face of the slab are shown in the figure. In Fig. 4.6a, the ray will be refracted once in the interface between the air and the slab. It will then pass through the

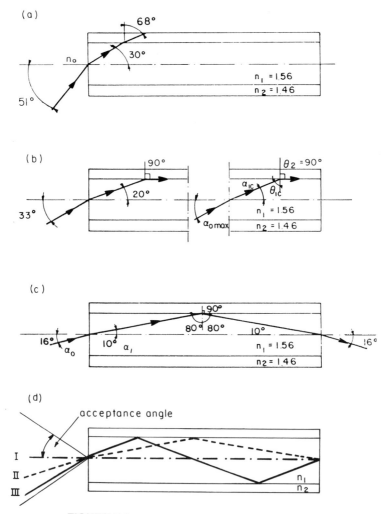

FIGURE 4.6   Total internal reflection in a slab waveguide.

inner layer and will be refracted into the outer layer. This will occur when the angle of incidence in the air is approximately 51°, as shown in the figure. In Fig. 4.6b, the ray is incident on the surface at an angle 33° and will be refracted once in the inner layer and then refracted again to the interface between the inner layer and outer layer. In Fig. 4.6c the ray, which has an angle of incidence smaller than 33°, will be totally internally reflected into the inner layer. It will move inside the inner layer of the slab to the other side and will again be totally internally reflected. In a series of internal reflections, the beam will thus be sent to the other end of the slab and emerge from that end, as shown.

Figure 4.6c illustrates that the angles of reflection in the inner layer are always identical. As the ray exits the inner layer, it is an exact mirror image of the path of the incident ray. The angle of the output beam is therefore equal to the angle of the input beam. For different incident angles, one obtains different zigzag paths inside the inner layer, as illustrated for rays I, II, and III in Fig. 4.6d.

4.4.3 Acceptance Angle and Numerical Aperture

Elementary geometry can be used to calculate which rays will be totally internally reflected (Wilson and Hawkes, 1989). Assume that the angle of incidence from air into the center of the slab is α_0, as shown in Fig. 4.6. Inside medium 1, the angle of the refracted beam (with respect to the normal to the interface between the media) is α_1, also shown [10] in Fig. 4.8. By Snell's law,

$$n_0 \sin \alpha_0 = n_1 \sin \alpha_1. \tag{4.3}$$

Geometrical relations give

$$\sin \alpha_1 = \cos \Theta_1 = (1 - \sin^2 \Theta_1)^{1/2} \tag{4.4}$$

and therefore

$$n_0 \sin \alpha_0 = n_1 (1 - \sin^2 \Theta_1)^{1/2}. \tag{4.5}$$

The angle α_0 can assume many values, but in order to be have total internal reflection from the surface between medium 1 and medium 2, the maximum value of α_0 is α_{0max}. This corresponds to the value Θ_{1c} and to $\Theta_2 = 90°$ as shown in Fig. 4.6b. For α_0 larger than this critical value, the beam will be refracted into medium 2 and not totally internally reflected. At the critical value,

$$n_0 \sin \alpha_{0max} = n_1 (1 - \sin^2 \Theta_{1c})^{1/2}. \tag{4.6}$$

By substituting the value n_2/n_1 for $\sin \Theta_{1c}$ [see Eq. (4.2)] it is possible to obtain the angle α_{0max}, which is the value of α_0 corresponding to Θ_{1c}. The angle α_{0max} is the maximum incidence angle for which total internal reflection in the slab is obtained, as shown in Fig. 4.6b. All the rays of light incident at an angle $\alpha < \alpha_{0max}$ will be totally reflected inside the slab.

---

[10] Figure 4.8 illustrates one of the cross sections of a cylindrical optical fiber. This cross section is identical to the one shown in Fig. 4.6. The symbols and the formulas used here apply to both figures.

The literature on fiberoptics generally does not mention the value of α_{0max}; instead, it concentrates on the parameter $n_0 \sin \alpha_{0max}$, which is called the numerical aperture (NA). In addition to being a useful quantity when considering the optical properties of lenses, NA is a good measure of the light-gathering power of optical fiber components. The following section includes simple calculations of NA; its full significance is clarified in later sections.

It is not difficult to show (Wilson and Hawkes, 1989) that NA is given by the expression

$$\text{NA} = n_0 \sin \alpha_{0max} = n_1[1 - (n_2/n_1)^2]^{1/2} = (n_1^2 - n_2^2)^{1/2}. \quad (4.7)$$

One could also write an approximate expression for the numerical aperture:

$$\text{NA} = n_0 \sin \alpha_{0max} \approx n_1(2\Delta)^{1/2} \quad (4.8)$$

$$\Delta = (n_1 - n_2)/n_1 = (1 - n_2/n_1). \quad (4.9)$$

EXAMPLE (i): In Fig. 4.6, $n_1 = 1.56$ and $n_2 = 1.46$; thus $\Delta = 1 - 1.46/1.56 = 0.064$. Therefore one can write NA $= (1.56^2 - 1.476^2)^{1/2} \approx 1.56(2 \times 0.064)^{1/2} = 0.55$, which corresponds to $\alpha_{max} = 33°$.

EXAMPLE (ii): Let us assume that medium 2 is soda lime glass with $n_2 = 1.52$ and medium 1 is flint glass with $n_1 = 1.67$; then $\Delta = 1 - 1.52/1.67 = 0.09$ and thus NA $\approx 1.67(2 \times 0.09)^{1/2} = 0.7$. This corresponds to $\alpha_{0max} = 43°$.

4.4.4 Transmission of Light through a Bent Slab Optical Guide

Substituting a slab whose shape is not straight, but half-circular (as shown in Fig. 4.7) into the simple geometrical calculations will prove that the conclusions just reached hold true even when the slab is shaped differently. Once again, total internal reflection occurs inside the inner layer and the output angle is equal to the input one. If the bending radius is too small, some of the light is not totally internally reflected and thus it emerges into the outer layer. This represents transmission loss in the slab.

4.4.5 Transmission of Light in Optical Fibers

The schematic drawings shown in Figs. 4.6 and 4.7 need not be flat slabs. Instead of a two-dimensional structure, it may also represent the cross section of a three-dimensional cylindrical structure. This structure consists of an inner rod (the central core) and an outer tube (the cladding). The core has a higher index of refraction than the cladding. The path of a beam of light in the three-dimensional rod again results in transmission by a series of total internal reflections. The explanation given above will hold for any cross section of this compound rod.

By reducing the physical dimensions of the glass rod until it becomes very thin, total internal reflection and light transmission will again occur as it becomes a thin thread of glass having both core and cladding. A fiber that consists of core

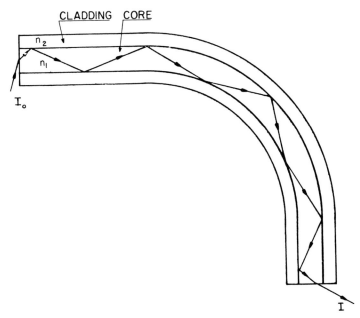

FIGURE 4.7 Total internal reflection in a bent slab.

and cladding is called a clad fiber. If it consists of core only (and the outer "cladding" is air), it is called an unclad fiber.

The calculations of the acceptance angle and numerical aperture derived for the two-dimensional slab can also be applied to the three-dimensional optical fiber. The main difference between the slab and the optical fiber is that the planar symmetry is replaced by cylindrical symmetry. Figure 4.8 represents one cross section of a cylinder; any ray of light impinging on the end with an angle $\alpha_0 <$ 33° will be transmitted through the rod (or the fiber) by total internal reflection.

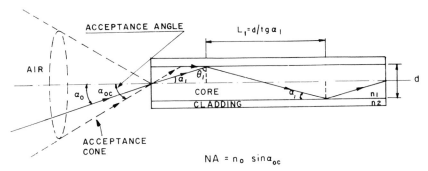

FIGURE 4.8 Light propagation in an optical fiber.

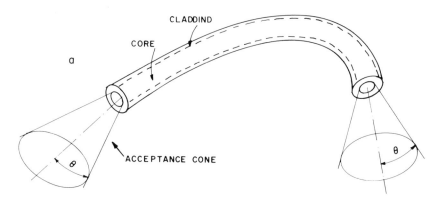

FIGURE 4.9 (a) Light transmission and the acceptance cone in an optical fiber. (b) Photograph of light transmission through optical fibers.

For a cylindrical rod, there are many cross sections that pass through the axis of the core, and all will result in an identical figure. There is therefore a whole cone of rays, all with the same angle, that behave identically. This holds true as long as the rays are transmitted through the rod or the fiber. If $\alpha_0 > 33°$, the rays will be refracted into both the cladding layer and the air. The angle of the cone is called the acceptance angle of the rod or the fiber. A schematic drawing of this acceptance cone is shown in Fig. 4.9. As in the case of a slab guide, the important concept of numerical aperture will be used, as given in Eq. (4.7).

4.4.6 Path Length in a Waveguide

Let us now consider a ray of light which is incident on a slab waveguide with an incident angle α_{01}, as shown in Figs. 4.6d and 4.8. The ray is totally reflected several times before it emerges from the slab. If the width of the inner layer of the slab is d, the distance between two consecutive reflections is given by $L_1 = d/\tan \alpha$. If the total length of the slab is L, the number of total internal reflections in the slab, N_1, is given by

$$N_1 = \frac{L}{L_1} = \frac{L}{d} \tan \alpha.$$

For a larger angle of incidence α_{02}, a larger α_2 will be obtained and the number of internal reflections N_2 will be increased, as shown in Fig. 4.6d. This general conclusion applies to both slab waveguides and optical fibers, as discussed in the following.

4.5 OPTICAL PROPERTIES OF OPTICAL FIBERS—ADVANCES

Optical fibers tend to be fairly rigid. If a truly flexible fiber is necessary, its outer diameter must be smaller than 0.1 mm. The actual diameter of the core and the cladding may vary with different fibers. First under discussion is a simple example, where the core diameter is 0.01 mm, $n_1 = 1.62$, and $n_2 = 1.52$. Several simple observations may be made (Allard, 1990; Cherin, 1983):

Input versus Output in an Optical Fiber

Figure 4.8 represents one of the many cross sections of the fiber. It is possible to view the figure as rotating along a horizontal line located at the center of the fiber. The result is a full three-dimensional picture of ray propagation in a fiber. In any event, all the rays that are incident on the input face of the fiber can be thought of as being located on a surface of a cone whose apex angle is α_0. The output will also consist of rays that are on a surface of a cone, having the same apex angle α_0.

Input and Output for a Fiber Immersed in Liquid

There is a marked difference in the NA values if the fiber end is immersed in air or in water, as illustrated in the following cases:

EXAMPLE (i)—*Fiber immersed in air*: Indices of refraction: core $n_1 = 1.62$, cladding $n_2 = 1.52$, immersion medium $n_0 = 1.00$ (air). Using (4.8), we obtain

$$\text{NA} = n_0 \sin \alpha_{0\max} \approx n_1 (2\Delta)^{1/2} = 0.56.$$

corresponding to $\alpha_{0\max} = 34°$. This is the maximum angle for both the input and the output of the fiber.

EXAMPLE (ii)—*Fiber immersed in water*: For the immersed water medium, $n_0 = 1.33$. The same calculation again yields $\text{NA} = n_0 \sin \alpha_{0\max} = 0.56$, but with $n_0 = 1.33$ a new value $\sin \alpha_{0\max} = 0.42$ is obtained; therefore $\alpha_{0\max} = 24.8°$.

EXAMPLE (iii)—*The input end of a fiber is in air and the output is immersed in water*: At the input end the maximum input angle is $\alpha_{0\max} = 34°$, as in (i). At the output end the maximum output angle is reduced to $\alpha_{0\max} = 24.8°$, as in (ii).

Input versus Output for a Focused Beam of Light

Suppose that a laser beam is focused by a lens on the proximal face of a fiber. The input beam can be viewed as consisting of many individual rays, each of which has a different input angle α (with respect to the normal). Assuming that $0 < \alpha < \alpha_{\max}$, where $\alpha_{\max}$ is the maximum input angle mentioned in Eq. (4.6), the cone of light that can be transmitted through the fiber has a maximal apex angle that is determined by the NA.

For each of the angles α, one may repeat the argument given above in (ii). The output beam will also consist of many individual rays that exit the fiber with angles α identical to the input angles. If the beam that impinges on the input face of the fiber is a solid cone of light whose apex angle is $\alpha_{\max}$, the output beam will therefore also consist of a solid cone of apex angle $\alpha_{\max}$, as shown in Fig. 4.9.

Bent Fiber

The main purpose of the optical fiber is to conduct light regardless of whether the fiber is straight or flexed. If the half-circular slab is replaced by a cylindrical optical fiber, the results will be the same; total internal reflection through the fiber will be obtained and the input cone of acceptance will be equal to the output cone. This is shown in Fig. 4.10.

Skew Rays

Until now, only rays of light that are limited to one meridional plane have been considered. While also taking into account the rays that are totally internally reflected in this plane, only one of the many cross sections of the optical fiber was considered. This applies when the input beam itself is in this plane. In reality, however, other options exist which make it possible to send, through fibers, rays that are tilted with respect to this plane. These are called skew rays because their path is not constrained to one plane. They bounce back and forth while being totally reflected from the core–clad interface; their path is three-dimensional. The treatment of such rays is rather complicated and will not be dealt with here. The main conclusions reached earlier, however, are applicable even when skew rays are considered.

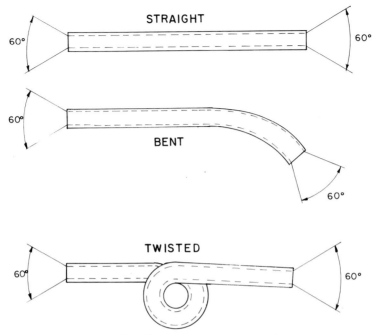

FIGURE 4.10 Optical properties of bent fibers.

Path Length inside an Optical Fiber

After all these internal reflections, what is the actual path length made by the ray of light through a given length of fiber? Within the confines of the meridional plane of a fiber, the situation is identical to that of a flat slab. By following the path of a ray inside the core, as in Fig. 4.8, it is seen that between two consecutive reflections the beam travels a distance $L = d/\tan \alpha_1 = d \tan \Theta_1$. Both the total number of reflections N and the total path length in a fiber whose length is K can be calculated from this relation: $N = K/L$.

EXAMPLE: Let us assume that $\Theta_1 = 80°$ and that $d = 0.1$ mm. The distance between consecutive reflections is $L = 0.1 \tan 80 = 0.5$ mm in the horizontal direction. With a fiber whose length is $K = 1$ m, the total number of reflections would be 1000 mm/0.5 mm = 2000.

Intensity Decrease Due to Reflections

Suppose that the intensity of the incident beam inside the core is $I = 1$ and the reflectance is 0.999998. After one reflection the intensity will be 0.999998, and after two reflections it will be $(0.999998)^2$. After the 2000 reflections mentioned above, the intensity will be only slightly reduced to $(0.999998)^{2000} = 0.996$. If instead of total internal reflection there was a metallic mirror with $R = 0.95$, the same calculation for only 100 reflections would give $(0.95)^{100} = 0.006$. Hence, the high reflectance in an optical fiber is quite important for light transmission.

Various Launching Conditions

Ray III in Fig. 4.6d is an extreme case in which the angle of reflection inside the core is close the critical angle. For this ray the number of reflections is maximal. Rays which are closer to the fiber axis will be transmitted with reduced loss, for two reasons. First, for these rays there is a lower number of reflections and the reflection losses mentioned above are decreased. Second, the path length of the beam in the core is smaller, leading to reduced absorption losses.

Numerical Aperture

The maximum cone of transmittable light (which is given by the NA) is determined by the core diameter and the indices of refraction n_1 and n_2 of the core and the cladding. Both a large diameter and a significant difference $n_1 - n_2$ result in a wide acceptance angle. If the difference $n_1 - n_2$ is small and the core is small, the opposite result occurs. It can be shown that $(NA)^2$ determines the ability of a fiber to collect light from an extended source, such as from a lamp. Some values of NA and of the corresponding acceptance angles, β, are given in Table 4.1.

Bending Losses

For most practical applications, fibers are used in a flexed position. Geometrical calculations in this case are somewhat more complicated, but the main results are the same. With a fiber bent in a circle of diameter a, the incident cone is identical to the exit cone. During the propagation of light in the bent fiber, additional losses occur because of bending. The losses are determined by n and d.

Modes in Fibers

In regular optical fibers, light may propagate in several "modes" (not to be confused with the laser modes). In each mode, the beam makes a different zigzag path inside the fiber core. For any given fiber, one can calculate the total number of modes N in the fiber. Let us consider a fiber of core diameter d and indices of

---

TABLE 4.1   **Numerical Aperture and Acceptance Angles in Fibers**

n_1		n_2			
		1.46	**1.48**	**1.50**	**1.52**
1.56	NA	0.550	0.493	0.428	0.351
	β (deg)	33.3	29.5	25.4	20.5
1.58	NA	0.604	0.553	0.496	0.431
	β (deg)	37.2	33.6	29.8	25.5
1.60	NA	0.655	0.608	0.557	0.500
	β (deg)	40.9	37.4	33.8	30.0
1.62	NA	0.702	0.659	0.612	0.560
	β (deg)	44.6	41.2	37.7	34.1

refraction n_1 and n_2 of the core and cladding. The total number of modes (Wilson and Hawkes, 1989) is related to the variable V, which is given by

$$V = \pi d/\lambda (n_1^2 - n_2^2)^{1/2} \approx \pi d/\lambda n_1 (2\Delta n_1)^{1/2}. \qquad (4.10)$$

For large values of V, the total number of modes N in the fiber is given by

$$N = V^2/2, \qquad (4.11)$$

and it can be approximated by the expression:

$$N = (d\pi/\lambda)^2 n_1^2 \Delta. \qquad (4.12)$$

Single-Mode Fibers

There are special cases, such as when the core diameter is very small, in which only one mode can propagate in the fiber. These fibers are called single-mode fibers. It can also be shown that if $V < 2.4$, the fiber is a single-mode fiber. Single-mode fibers will be quite important in optical communications. There is also interest in fiber bundles consisting of single-mode fibers for imaging purposes. Such bundles will have very high resolution and may also be useful for holographic applications.

EXAMPLE: If $d = 50$ μm, $n_1 = 1.53$, $n_2 = 1.50$, and $\lambda = 0.5$ μm, then $V = 50$ and the total number of modes in the fiber is $N = 2500$. If the core diameter is reduced to $d = 2.5$ μm then $V < 2.4$ and the fiber is a single-mode fibers.

4.6 THE FABRICATION OF OPTICAL FIBERS—PRINCIPLES

In principle, but not in practice, the method for preparing thin optical fibers is quite simple. One end of a glass rod is heated to the point where the glass is soft and malleable. The end is held with a special tool. While constant heat is supplied, the tool is pulled away and a thin glass thread (fiber) is formed.

Light can be transmitted in an unclad fiber. Its total internal reflection takes place at the interface between the fiber and air. However the surface of an unclad fiber is exposed to moisture, dirt, and mechanical damage such as scratches. In each internal reflection, the light may therefore either be scattered outside the fiber by the scratches or be absorbed by the dirt. This will severely limit the transmission through the fiber. Thus, fibers which consist of core and cladding are better. The internal reflections inside the fiber take place at the interface between core and cladding and the cladding layer prevents the deleterious effects mentioned above. The core–cladding interface can have much better optical qualities than the outer surface of an unclad fiber. Most of the fibers that are in use today are therefore clad fibers.

The pulling technique can be modified to fabricate clad optical fibers. This technique requires a piece of glass, known as a preform, that consists of two components—an outer region of lower index of refraction and an inner rod with a

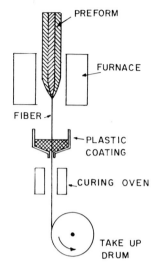

FIGURE 4.11   Fabrication of optical fibers.

higher index of refraction. This is shown schematically in Fig. 4.11. One end of the preform is heated until it becomes soft. This end is pulled away from the preform, thus forming a thin fiber that consists of core and cladding.

In practice, the fabrication of good optical fibers is complex. It involves several important considerations (Dreyfus, 1986; Lines, 1984):

(i) Preparation of pure starting material: A major fraction of the losses in optical fibers stems from impurities present in the preform glass from which the fiber was made. For example, 1 part per billion of impurities such as iron (Fe^{3+}), chromium (Cr^{3+}), or other transition metal ions causes noticeable absorption of visible light. Ten parts per billion of hydroxyl (OH^-) ions give rise to strong absorption at 2.7, 1.39, 1.25, and 0.95 μm. Impurities are especially troublesome when the fibers are used for power transmission (e.g., laser surgery). Methods have been developed to purify the materials to a high level.

(ii) Uniform pulling: In order to obtain a long fiber of uniform cross section, the thickness of the fiber is controlled by regulating the pulling speed.

(iii) Preform: The compound preform is prepared and requires careful selection of materials to be used for the core and cladding. These two materials must be intimately connected when the preform is heated and a clad fiber is pulled.

Two families of materials are commonly used:

- High-silica glasses. These are based on pure SiO_2 (fused silica), which has a low index of refraction (i.e., 1.45). The index of refraction can be increased during the preparation of the fused silica by incorporating a few percent of dopant material such as GeO_2 or P_2O_5. The index of refraction can also be decreased by incorporating dopants such as Be_2O_3. All these

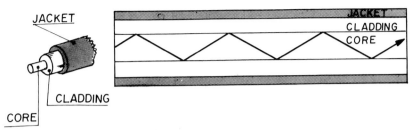

FIGURE 4.12 Fiber with core, cladding, and jacket.

materials have a fairly high melting point of approximately 2000°C.
- Silicate glasses. These are the classical glasses, such as the sodium boro-silicate glass ($Na_2O–B_2O_3–SiO_2$) or soda lime glass ($Na_2O–CaO–SiO_2$). They have a much lower melting point (800–1000° C).

The preform may be prepared by depositing the core glass, of higher index of refraction (e.g., high silica glass), inside a tube made of glass having a lower index of refraction (e.g., fused silica). Alternatively, a glass having a lower index of refraction is deposited on a rod made of glass of a high index of refraction.

(iv) Plastic jacket: It is occasionally necessary to protect the clad fiber after it is made. The mechanical properties of this thin thread of glass may not be adequate for some applications, and it may break if flexed. Exposing the cladding layer to the chemical effects of water or solvents or to physical effects, such as scratching, may cause problems. In order to protect the whole clad fiber, it is normal to apply an outer layer of plastic material. The fiber passes through the container holding the liquid plastic at the same time that the fiber is being pulled. A thin and uniform coating of plastic material is formed around the fiber. This layer may then be hardened by exposure to UV radiation or by heat, forming a outer plastic jacket which protects the fiber. A cross section of a clad fiber with a plastic jacket is shown in Fig. 4.12.

It is interesting to note that Charles Boys, who used a crossbow and arrow, made his fibers at an approximate rate of 30 m/sec (i.e., 70 mph). Modern fabrication methods are not so different, as the pulling rate of the fibers is about 40–50 m/sec (120 mph).

4.7 SPECIAL OPTICAL FIBERS FOR UV, VISIBLE, AND IR LIGHT—PRINCIPLES

Fibers used for optical communications are designed to transmit signals that are emitted from small lasers or light-emitting diodes. The main requirement is that transmission losses will be extremely small so that the signal can be transmitted over great distances. Currently, communications scientists and engineers have decided to use optical signals in the visible (0.4–0.7 μm) or near IR (0.8–1.5 μm) parts of the optical spectrum. The fibers used for these applications are based on glasses that transmit well in this spectral region. Since the power emitted by the tiny diodes used in communications is small, the "power transmission" demands on the fibers are also modest. The material used for these purposes is ordinary silicate glass, much like window glass, but of high quality, that is, high purity and low scatter. Optical communication fibers are similar to those that will be described for image transmission or sensors.

For many applications, optical fibers made of regular silica glass are insufficient and different types of fibers are needed. For example, these fibers are not the best choice for the transmission of extremely high powers, and pure fused silica or high-silica fibers serve this purpose better. The transmission of optical radiation of different wavelengths through an ordinary glass fiber is limited by the spectral transmission of the glass. They are opaque in the UV region, at wavelengths below 300 nm, and in the infrared region above 2.5 μm. Different materials are needed for the transmission of these wavelengths, and these special fibers are the topic of this section. The special materials used for the fabrication of such fibers are listed in Fig. 4.13.

Earlier, we described the geometrical optics model for total internal reflection from the interface between core and cladding, neglecting details of the interface phenomena. The "true" physical optics picture is slightly different. Although the transmission of energy in the fiber occurs mostly in the core, a small percentage of the energy is transmitted via the cladding. If the cladding layer is not highly transparent at the transmitted wavelength, the transmission of the whole fiber decreases. Thus, for a UV-transmitting fiber, the cladding, as well as the core, must be highly transparent in the UV. Similarly, for IR-transmitting fibers, both the core and the cladding must be transparent in the IR.

Again, it is quite surprising that even in the development of "special" fibers, nature was first. There are reports that the individual hairs of Arctic mammals, such as the polar bear, resemble fused silica ("quartz") fibers. There are indications that the hairs are exposed to UV radiation from the sun and transmit it to the skin, where it is converted to heat. This is a natural fiberoptic solar collector that helps to keep the animals warm. IR-transmitting fibers also occur in nature and are possibly used for optical communication between moths (see Chapter 7).

4.7.1 Special Fibers for Visible Radiation

Optical fibers from silica-based glasses have several properties that make them extremely useful. They are the standard fibers that are used for both communications and medical applications. Over the years, there has also been increased interest in using plastic materials for the manufacture of optical fibers. Some of the earliest experiments with light transmission in optical fibers were

OPTICAL TRANSMISSION OF MATERIALS

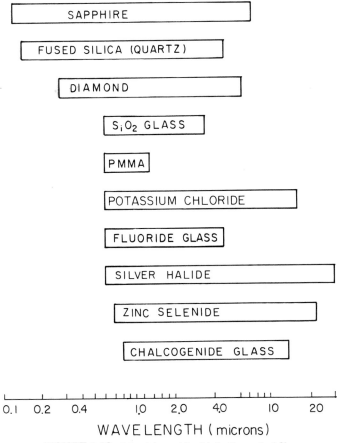

FIGURE 4.13 Materials used for fabricating optical fibers.

performed using plastics. They were, however, of poor quality and the rapid development of satisfactory glass fibers superseded them.

There are standard ways in which polymer scientists have made fibers from a large number of polymers. Similar methods were applied for the fabrication of optical fibers. For example, there has been a growing demand by the automobile industry for inexpensive optical fibers. The optical properties of these fibers have been improved and are still undergoing development. Some of these plastic fibers may eventually be used in medicine. Although the price of good-quality glass fibers has dropped tremendously during the last decade, some applications such as disposable optical fibers for diagnostic devices require even cheaper fibers.

When the difference in optical quality between glass fibers and plastic fibers is unimportant, less expensive plastic fibers may become the fibers of choice.

4.7.2 Ultraviolet-Transmitting Optical Fibers

Ultraviolet laser light needs to be transmitted through fibers in two types of applications:

(1) Low-power applications, such as the spectroscopic identification of tissue. HeCd, Ar, or Kr ion lasers are typically used.

(2) High-power applications such as tissue ablation during endoscopic laser surgery. Excimer lasers are typically used.

In principle, UV-transmitting fibers can be fabricated from materials that are transparent to UV light, such as LiF or CaF_2 crystals. Efforts to fabricate optical fibers from these hard crystals have not yet resulted in useful fibers. One of the few materials that has been used successfully for UV fibers fabrication is SiO_2, amorphous or crystalline, that is transparent between 200 nm and 4 μm. Amorphous glasses made of pure SiO_2 are called fused silica or vitreous silica and can be drawn into fibers. These fused silica fibers are sometime called quartz glass fibers, or quartz fibers, although strictly speaking the word quartz refers to the single crystal. The fabrication of these fibers is similar to the fabrication of the ordinary glass fibers discussed earlier.

The main problem with fused silica is its low index of refraction ($n = 1.46$ at $\lambda = 0.5$ μm), which makes it quite difficult to find a suitable material for cladding. Because of the difficulties in finding suitable glass for the cladding, attention was turned to the use of polymers. For example, a thin layer of plastic such as soft silicone resin or hard fluoropolymer may be used as a cladding layer. It has been applied as a sheath on the fused silica fiber by passing it through a bath containing the resin. The fibers are called soft clad silica and hard clad silica fibers. The resin cladding of UV-transmitting fibers does not transmit well in the UV and therefore the total transmission of the clad fiber in the UV range is decreased. A newer technique is based on pure silica for the core and on silica which contains impurities of phosphorus (called P doped) for cladding with lower refractive index. The glass-clad silica fibers are then coated with silicone resin to improve their mechanical properties. These fibers are more difficult to make but have better transmission properties. For most applications in the ultraviolet, fused silica fibers are a good choice. They transmit well in the spectral range of $\lambda > 300$ nm and are thus useful at the excimer wavelength 308 nm. The cross sections of several useful fused silica fibers (McCann, 1992) are shown in Fig. 4.14.

Much research is currently being done on other materials that transmit well in the middle and far UV and can be drawn into fibers. Such fibers would be invaluable for the transmission of excimer laser beams, including wavelengths shorter than 300 nm, which may also have clinical applications.

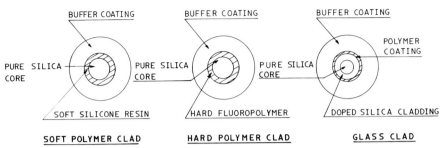

FIGURE 4.14 Fused silica fibers.

4.7.3 Infrared-Transmitting Optical Fibers

The infrared spectrum can be divided into three arbitrary ranges. These, and the lasers that fall in each range, are noted:

(i) Near infrared (NIR), $0.7-1.5$ μm; laser: Nd:YAG
(ii) Mid-infrared (MIR), $1.5-9$ μm; lasers: Er:YAG, Ho:YAG, HF, CO
(iii) Far infrared (FIR), >9 μm; laser: CO_2

There are three families of materials that have been considered for the transmission of radiation in these regions (Drexhage and Moynihan, 1988; Harrington, 1990).

(a) Hollow waveguides: These are flexible hollow tubes. Some are made of infrared-transmitting glasses; others are made of hollow glass or plastic tubes with their inner surface coated with a metallic layer. Several have an inner surface coated with a thin layer of a transparent material (e.g., dielectric). Although these waveguides transmit well in the whole infrared range, there is an increase in power loss when the fiber is bent and such waveguides have limited mechanical flexibility.

(b) Infrared-transmitting glasses: Regular fused silica fibers are suitable for the transmission of near-infrared radiation and have been widely used for the transmission of Nd:YAG laser light. Fused silica fibers, with low OH content, are suitable for the transmission of laser radiation up to 2.2 μm (e.g., holmium laser). For the transmission of mid-IR, special glasses such as those made of ZrF_4 or other fluorides are necessary. These fluorides are not transparent in the far infrared and thus other infrared-transmitting glasses are considered, such as As_2S, or other chalcogenide glasses. Many of these new glasses suffer from poor mechanical properties, high toxicity, and solubility in water, making them poor candidates for IR fibers.

(c) Crystalline fibers: Many crystalline solids are highly transparent in the infrared and a few may be useful for fabricating fibers. Some are made by pulling from crystalline melt, like regular glass fibers, and others by methods such as extrusion. Fibers made from certain crystalline halide materials are flexible and highly transparent in the middle and far infrared. They are the only thin and flexible fibers which have been used successfully for transmission of the high-power CO_2 laser beam.

There has been an enormous effort in developing glass fibers for communications and these fibers have now almost reached the theoretical limits of light transmission. Fibers for the UV and IR are less well developed. Losses are significantly higher and the mechanical and chemical properties are certainly inferior. Nevertheless, with increase in demand and the development of new laser–fiber systems, there is no doubt that UV- and IR-transmitting fibers will also improve.

4.8 POWER TRANSMISSION THROUGH OPTICAL FIBERS—PRINCIPLES

The problem of transmitting high power through optical windows and lenses has been studied by scientists and engineers interested in high-power lasers, such as for controlled fusion or "star wars" technology. Extremely high power densities must be transmitted through the windows. Two types of absorption mechanisms cause serious problems:

(i) Extrinsic absorption: Just a speck of dirt or moisture, or a scratch on the window's surface, or slight absorption due to imperfections inside the window may cause severe damage.

(ii) Intrinsic absorption: Other more complex mechanisms also can lead to optical damage in windows at high power densities. These may appear even when extrinsic absorption is absent. Some (e.g., laser-induced breakdown) are mentioned in Section 3.4.5.

The solution to these problems is relevant to power transmission in optical fibers. Some of the loss mechanisms in optical fibers that manifest themselves at high laser power levels are shown in Fig. 4.15.

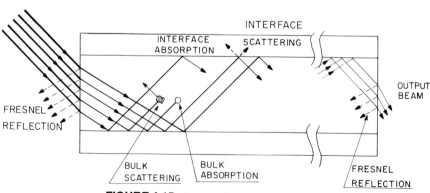

FIGURE 4.15 Loss mechanisms in optical fibers.

Other problems are distinctively different from those of high-power windows and involve the coupling a laser beam that propagates freely in the air into an optical fiber. This coupling is possible with the help of a focusing lens or a mirror and is a fairly complex problem (Schonborn *et al.*, 1986; Kar *et al.*, 1989).

As every physician or scientist who tries to use fiber–laser systems for therapeutic applications knows, laser power transmission through fibers is quite complicated. At present, both commercial and research systems are in a development stage. Some of the physical problems to be considered in the design of a laser–fiber system are now presented. Some of the more practical aspects of this problem are discussed in Section 8.4.

The simple system consists of a laser, an optical fiber for transmitting the laser power ("power fiber"), and a lens for coupling the laser light into the fiber. Such a system is shown schematically in Fig. 4.16.

Each fiber has an acceptance angle (see Section 4.5) and all rays of light that impinge on the input end of the fiber at a smaller angle can be propagated through

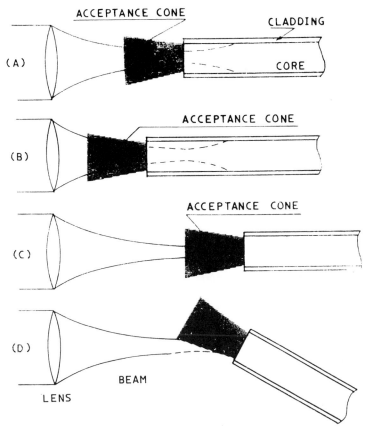

FIGURE 4.16 Coupling of laser light into a "power" fiber.

the fiber. A laser beam whose diameter is often several millimeters has to be coupled into the core of a fiber whose diameter is often a fraction of a millimeter. For this purpose, one could use short, medium, or long focal length lenses. Use of the short focal length lens presents two problems: (a) the angle subtended by the lens at the fiber input may be higher than the acceptance angle and some of the laser energy will not be coupled into the core; (b) at the focal point of the lens, the spot size is very small (see Section 2.2.3), much smaller than the fiber core diameter. The laser power is concentrated at the focal spot and the power density may be so high that the fiber will be damaged. With a long focal length lens the angle subtended at the fiber input is smaller. The long focal length lens will have a large focal spot (see Section 2.2.3) so that the power density at the fiber tip is much smaller. On the other hand, if the focal spot is equal to or larger than the core diameter, some of the energy will be directed to the cladding layer and will be lost. One solution is to use a medium focal length lens, as shown in Fig. 4.16. The focal spot may be just inside the fiber, as shown in Fig. 4.16b. If the power density there is still too high and laser-induced breakdown occurs, the beam may be focused just outside the fiber, as shown in Fig. 4.16c. The fiber axis must obviously be parallel to the laser beam to prevent a situation such as shown in Fig. 4.16d.

Examples of the powers which have been transmitted by optical fibers with lengths of few meters are as follows (see also Chapter 8):

   (a) Nd:YAG: 1000 W was transmitted continuously through fused silica fibers with a diameter of less than 1 mm.
   (b) CO_2: 100 W was transmitted continuously through flexible crystalline fibers with a diameter of 1 mm and more than 1000 W through hollow dielectric waveguides.
   (c) Excimer: energy pulses of 1.5 J (pulse length 70 nsec; peak power 20 MW) were transmitted through fused silica fibers with a diameter of 1 mm.
   (d) Er:YAG: Pulses of 2 J (pulse length 20 µsec; peak power 10 kW) were transmitted through fluoride fibers with a diameter of 0.6 mm.

4.9 MODIFIED FIBER ENDS AND TIPS—PRINCIPLES

The input and/or the output ends of fibers can be modified in a number of ways. Both ends can be cut and polished at an angle to the fiber axis, or they can be heated and shaped into a lens, a ball, or a taper. These shapes have potential applications for fiber tips, as discussed in Chapter 8. The paths of beams of light in a tapered fiber input or output are shown in Fig. 4.17. The output beam from tapered output ends is less diverging and may be useful if a more collimated beam is necessary. On the other hand, a tapered input end is useful for high power delivery. If one uses a laser beam of high power but with a large cross section, the

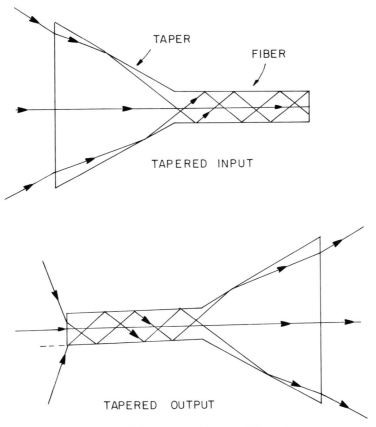

FIGURE 4.17 Beam path in tapered fiber ends.

power density at the tapered end is relatively low. The beam will be coupled through the tapered section into the fiber. The advantage of this scheme is that the power density at the input is low, which decreases the likelihood of breakdown due to excessive power density.

The paths of laser beams in a conical (pointed) and in a hemispherical fiber tip are shown schematically in Fig. 4.18. In a conical or an angled tip, the exiting beam may be sent sideways, which is useful in some applications. In a spherical or ball-shaped tip, the beam may be focused at a spot; this may help in fiberoptic laser surgery. Far from the focal spot, the beam diverges; this may be useful in cases in which lower power densities are needed, such as laser heating. Photographs of a ball-shaped tip and of a laser beam focused by this tip are shown in Fig. 4.19. Further discussion in given in Chapter 8.

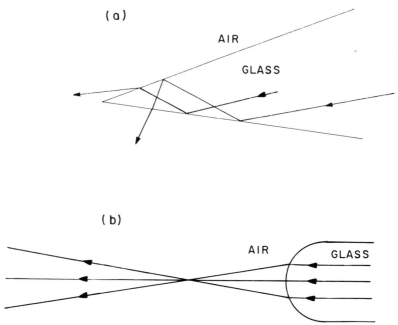

FIGURE 4.18   Schematic drawing of beam paths in (a) conical and (b) hemispherical fiber tips.

4.10 FIBER LASERS—ADVANCES

Until now, fibers have been discussed as passive components for the transmission of optical radiation, whether for illumination or power transmission. Other fibers serve different purposes. Chapter 2 mentioned that all lasers have an active medium in which light is amplified. For example, the active medium of solid lasers consists of a rod of solid material that provides gain. Nd: YAG and Er: YAG lasers are based on single crystalline rods of YAG that are doped with Nd or Er ions. There are similar lasers which are based on rods of glass doped with Er or Nd. There is no reason why this rod cannot be drawn to form a fiber, allowing the lasing action to take place inside the fiber itself. Actually, the operation of an Nd: glass laser was demonstrated first in optical fibers made of Ba crown glass doped with Nd ions (Snitzer, 1961). The fiber was pumped by a flash lamp and lasing occurred at a wavelength of 1.06 μm, similar to Nd: YAG lasers.

The idea of a practical fiber laser lay dormant for two decades. During the last 5 years there has been a renewed interest in the problem. Various fiber lasers have been studied and most of them were based on fibers made of silica glasses and on IR glasses (France, 1991). At this stage of development they are not yet

FIGURE 4.19 Photographs of (a) ball-shaped tip and (b) beam path in this tip. (Courtesy of R. Verdaasdonk.)

suitable for medical applications. The concept is intriguing, because in principle a fiber can be inserted inside an endoscope or a catheter and can be pumped until it lases. There are cases in which this fiber laser may offer advantages for endoscopic laser surgery, tissue welding, or ablation. A fiber laser is shown in Fig. 4.20.

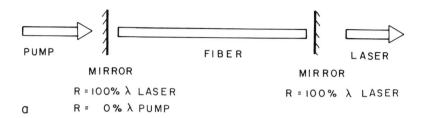

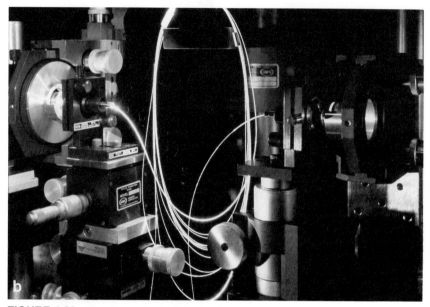

FIGURE 4.20 Fiber laser: (a) schematic drawing and (b) photograph. (Photo courtesy of GTE Laboratories, Inc., Waltham, MA 02254.)

References

Allan, W. B. (1973). *Fiber Optics: Theory and Practice*. New York: Plenum Publishing.

Allard, F. C. (1990). *Fiber Optics Handbook*. New York: McGraw-Hill.

Cherin, A. H. (1983). *An Introduction to Optical Fibers*. New York: McGraw-Hill.

Curtiss, L. E., Hirschowitz, B. I., and Peters, C. W. (1957). A long fiberscope for internal medical examinations. *J. Opt. Soc. Am.* **47**, 117.

Drexhage, M. G., and Moynihan, C. T. (1988). Infrared optical fibers. *Sci. Am.* **259**, 76–81.

Dreyfus, M. G. (1986). Glass requirements in medical fiber optics. *Adv. Ceram. Mater.* **1**, 28–32.

France, P. W. (1991). *Optical Fibre Lasers and Amplifiers*. London: Blackie.

Harrington, J. A. (1990). *Infrared Fiber Optics*. Bellingham: SPIE Press.

Hecht, E. (1987). *Optics*, 2nd ed. Reading, MA: Addison-Wesley.

Hecht, J. (1985). Victorian experiments and optical communications. *IEEE Spectrum* **22**, 69–73.

Hopkins, H. H., and Kapany, N. S. (1954). A flexible fiberscope using static scanning. *Nature* **173**, 39–41.

Kao, K. C., and Hockham, G. A. (1966). Dielectric fibre surface waveguides for optical frequencies. *Proc. IEEE* **113**, 1151.

Kapany, N. S. (1960). Fiber optics. *Sci. Am.* **202**, 72–81.

Kapany, N. S. (1967). *Fiber Optics Principles and Applications*. New York: Academic Press.

Kar, H., Helfmann, J., Dorschel, K., Muller, G., Muller, O., Ringelhan, H., and Schaldach, B. (1989). Optimiaztion of the coupling of excimer laser radiation into Q-Q fibers ranging from 200–600 micrometer diameter. *Proc. SPIE* **1067**, 223–232.

Katzir, A. (1991). *Optical Fibers in Medicine VI*, SPIE Proceedings, Vol. 1420. Bellingham, WA: SPIE Press.

Katzir, A. (1992). *Optical Fibers in Medicine VII*, SPIE Proceedings, Vol. 1649. Bellingham, WA: SPIE Press.

Lines, M. E. (1984). The search for very low loss fiber optic materials. *Science* **226**, 663–668.

McCann, B. P. (1992). Silica core fibers for medical diagnosis and therapy. *Proc. SPIE* **1649**, 2–7.

Schonborn, K. H., Kobayashi, N., and Kersten, R. T. (1986). High power laser beam delivery systems in surgery: the technical aspect. *Proc. SPIE* **658**, 32–35.

Seippel, R. G. (1984). *Fiber Optics*. Reston VA: Prentice Hall.

Snitzer, E. (1961). Optical maser action of Nd in barium crown glass. *Phys. Rev. Lett.* **7**, 444–446.

van Heel, A. C. S. (1954). A new method of transporting optical images without aberrations. *Nature* **173**, 39.

Wilson, J. and Hawkes, J. F. B. (1989). *Optoelectronics: An Introduction*, 2nd ed. Englewood Cliffs, NJ: Prentice Hall.

Wolf, H. F. (1979). *Handbook of Fiber Optics: Theory and Aplications*. London: Granada.

5

Optical Fiber Bundles

5.1 INTRODUCTION

In the fourth millennium B.C., the technology of heating and pulling thin rods of glass was known in western Asia (Harden, 1969). Glassmakers also knew how to pull rods of glass of one color and assemble a "preformed cane" by fusing together several rods of different colors. This cane would then be heated and drawn to a small diameter. After cooling, the cane was cut into small discs. During the processes of pulling and fusing, the discs kept the color order of the initial rods! Many discs were then placed in a mold, carefully heated, and fused together to form bowls or other objects. This "glass mosaic" technique was also used by the Phoenicians who lived in Palestine 2000 years ago. Glass fibers were made in Europe during the Middle Ages and the technique is sometimes called *millefiori* after the Italian word for "one thousand flowers," referring to the beautiful decorations. Modern techniques for making bundles of fibers are similar to the technique used for glass mosaics.

Tens or hundreds of thin, flexible optical fibers that are inserted into a plastic tube form a fiber bundle. Bundles in which the individual fibers are not ordered are called light guides and are used for illumination. Bundles in which the fibers are systematically arranged are called ordered bundles and are used for image transmission. Both types of bundles made it possible to fabricate the early imaging fiberscopes (Hirschowitz *et al.*, 1958) and, later, the medical endoscopes (Kapany, 1967).

Bundles of cells that behave like optical fibers appear in nature. In plants, they serve for fiberoptic sensing, as discussed in Chapter 7. A striking example is image transmission in various types of animals' eyes. The oldest type of eye is the compound eye which is built of many rodlike cells called ommatidia (Hecht, 1987). Each ommatidium rod is a cell that processes a small segment of a picture, and it resembles an optical fiber with a small lens at its front end. In the compound eye, the picture is divided into small segments that are processed with the help of an ordered fiberoptic bundle. Compound eyes are common among insects, some of which (e.g., ants) have only tens of fibers, whereas others (e.g., flies) have thousands.

In other animals such as vertebrates and humans, the eye is different. A single lens system forms an image on a light-sensitive screen, the retina, and the image is converted into nerve pulses which are conducted to the brain. There are two types of receptors in the retina, rods and cones, each of which has a diameter of roughly $1-3$ μm. There are roughly 120 million rods and about 7 million cones that synapse with approximately 1 million nerve fibers in the optic nerve. The rods and cones have a higher index of refraction than the surrounding retina tissue, and they therefore behave like optical fibers (Enoch, 1967; Levi, 1980). Light incident on the rods and cones is transmitted through them to photosensitive material within them, leading to the generation of signals. The whole assembly of rods and cones is thus a bundle of optical fibers.

Illumination and image transmission through bundles of optical fibers (and endoscopes) are the topics of this chapter (Allan, 1973; Siegmund, 1978). Medical endoscopes that are based on these bundles are discussed in the following chapter.

5.2 NONORDERED FIBEROPTIC BUNDLES FOR LIGHT GUIDES—FUNDAMENTALS

The individual fibers in the bundles that are discussed in this section are not arranged in any ordered manner. The bundles are very flexible and useful for some specific illumination purposes. Often a physician needs to illuminate an area inside the body that is not easily accessible. Such illumination may be needed for imaging, diagnosis, or therapy. Optical fibers may thus be used to transmit the light from a light source to the required place. In ordinary light sources, such as incandescent lamps, light is emitted from a relatively large filament. As mentioned in Section 2.2, this type of light cannot be focused by a lens onto a small area. Light from an ordinary lamp therefore cannot be focused into the core of a thin optical fiber. A thick optical fiber (e.g., a fiber with an outer diameter >1 mm) is also impractical because it lacks flexibility. The solution is thus to make use of a bundle consisting of many thin fibers. This simple bundle has two advantages: flexibility and a large area. The individual fibers are glued to each other at their ends, and the rest of their length is free. Fibers are thus prevented from "escaping" the bundle; the whole bundle is still flexible and an outer plastic jacket is used to protect it.

Nonordered fiberoptic bundles or light guides are used for illumination in various medical instruments, including endoscopes, in which they illuminate areas inside the body. Intense light is necessary for the physician to see into the body and receive a good-quality image. This clear image allows the physician to make a good diagnosis and facilitates recording of the image with photographic equipment. In order to achieve this goal, high-intensity lamps such as high-pressure xenon, quartz halogen, or mercury lamps must be used. The light from such a lamp is focused onto the input end of the bundle by using a simple lens system, and light exiting the output end serves for illumination. Excessive heat generated in the input end due to the high intensity of the light can be a problem, however. Remedies for this problem include cooling the input end and filtering some of the IR light (which heats but does not illuminate) with a "heat-rejecting" optical filter.

5.3 NONORDERED FIBEROPTIC BUNDLES—PRINCIPLES

5.3.1 General Concepts

One of the most important properties of light guides is their ability to collect light from an extended light source (high-pressure mercury or xenon lamps) and transmit much of it into the body. The parameters that control this light transmission are described below.

Looking at the cross section of a fiberoptic bundle near the input end, an assembly of fibers is revealed. When light is focused on the end face of the whole bundle, each of the cores transmits light. The area occupied by the cores is thus used for light transmission, whereas the area occupied by the cladding transmits little light. If the overall diameter of each fiber is D and the core diameter is d, the ratio of the core area to that of the whole fiber is approximately $0.9(d^2/D^2)$. It pays to have individual fibers that have a large core diameter and a thin cladding layer (Morokuma, 1979).

Another consideration in designing the light-guiding fiberoptic bundle is the numerical aperture (NA) of the individual fiber. Intuitively, it is clear that in order to collect as much light as possible from a light source, the individual fibers in the bundle must have a large numerical aperture; a wide cone of light must be collected by each fiber. It was shown mathematically that the collection efficiency (i.e., the ability to collect light from a source) is actually proportional to $(NA)^2$; the higher the NA, the higher the light collection efficiency. The NA is determined both by d and by the difference between the indices of refractions of the core and the cladding [see Eq. (4.9)]. Bundle design takes this into account.

In the case of light guides, the individual fibers have a typical core diameter of $15-50$ μm and an outer diameter of $20-60$ μm. When trying to transmit high-intensity light, the input end often becomes hot and needs cooling. Glass materials are thus usually more appropriate than plastic.

In general, the unique optical properties of lasers make them unsuitable as light sources for imaging. When illuminated by laser light, a surface looks unevenly illuminated because of speckle. This phenomenon is caused by interference effects resulting from the temporal coherence of the light. One way to remove this speckle is to vibrate the light guide along its central part while keeping its ends fixed. This is, however, not a practical method.

5.3.2 Fabrication of Nonordered Fiberoptic Bundles

Single optical fibers are fabricated by drawing from a preform, as described in Section 4.6. The drawn fiber can be collected (but not aligned) on a rapidly rotating drum. The simplest way to fabricate a fiberoptic bundle is to collect a large number of turns on the drum. The assembly of fibers may be cut through and the ends can be glued together to produce a nonordered fiberoptic bundle. The two ends of the bundle are compacted in some mechanical device and bonded together with epoxy cement, or fused together. Both ends are ground and polished to give a good-quality optical finish. The input end is placed in a special ferrule which is inserted into the light source housing. The output end may also be mounted in a ferrule, or may be inserted inside an endoscope.

A more refined manufacturing method is to draw a large number of fibers simultaneously from a large number of preforms. All these fibers are collected as a bundle, and the whole bundle may now be wound on a drum. Sections of this bundle are then used for illumination purposes, as described above. This process is much more economical and results in longer bundles of better quality.

5.3.3 Special Nonordered Fiberoptic Bundles

Several nonordered bundles are unique and deserve special mention. Some of these complex bundles are used for beam shaping and others are made of special glasses.

5.3.3.1 UV Light Guides

Some applications (especially for diagnostic systems) require that an area be illuminated with UV light rather than with white light. A light beam falling on a particular sample will excite luminescence; the analysis of this luminescence gives information about the sample. Such an analysis can be carried out using optical fibers. The basic system consists of two light guides. Light emitted from a source is transmitted via a fiber (or a bundle of fibers) to the sample. A bundle of fused silica fibers is used in place of ordinary glass for the UV illumination because of its better transmission properties. Light reflected from the sample, or the luminescence light emitted by the sample, is collected by a second receiving fiber (or bundle) and transmitted to a detector. Often there is an optical filter or a monochromator in front of the detector, in order to obtain a spectral analysis of the received signal.

5.3.3.2 Beam-Shaping Bundles

Fiberoptic bundles can change the cross section of the light beam that enters them. Bundles consist of an array of independent fibers, each guiding light from

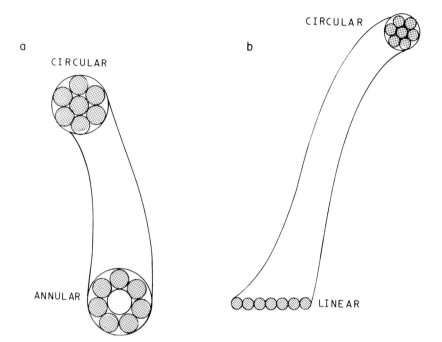

FIGURE 5.1 Nonordered bundles of fibers for beam shaping.

one end to the other. It is thus possible to have an array of fibers that is arranged in a full circle at the input end, with individual fibers at the output end arranged in a line, as shown in Fig. 5.1b, or in an open circle, as shown in Fig. 5.1a. Light from an extended source, such as an incandescent lamp, is easily focused on the circular input end of the bundle. The shape of the output beam has been changed by the fiber bundle. This may be important if the fibers are being used for illumination in medical instruments. The special arrangement of the output fibers helps in the mechanical design of the instruments and may result in a more uniform illumination of the field of view.

Both rigid and flexible endoscopes use optical fiber bundles for illumination. A fiberoptic illumination system for a rigid endoscope is discussed in Chapter 8 and is shown schematically in Fig. 8.5. The focusing lens, the heat-rejecting, filter and the light bundle inside the endoscope can all be seen.

Optical fiber bundles can also be used for beam splitting or beam combining. The two optical bundles shown in Fig. 5.2 can be combined into one bundle near a sample. Simultaneously, the two are still separated at the other ends: one near the source and one near the detector. This is possible because individual fibers can be grouped in a bifurcated array of fibers, called a Y guide; it is one example of a branched guide. It can be shown that if the NA of the individual fibers is large, the light collection efficiency of the Y guide is fairly high (i.e., the ratio between the

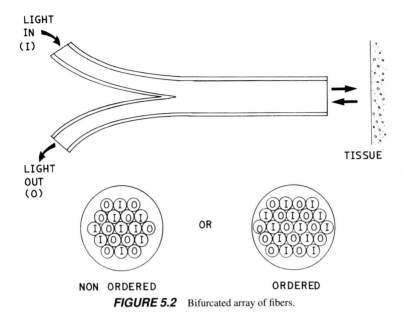

FIGURE 5.2 Bifurcated array of fibers.

input energy and the output energy is relatively high). The Y guide may be used to measure reflectivity or luminescence from tissue.

5.4 ORDERED FIBEROPTIC BUNDLES FOR IMAGING DEVICES—FUNDAMENTALS

Optical fibers can be accurately aligned in a bundle so that the order of the fibers in one end is identical to the order in the other end. Such an array of optical fibers is termed an ordered bundle. It is sometimes called a coherent bundle, but this term is not used here because it may be confused with the coherent nature of laser light. If an image is projected onto one end of a bundle, each individual fiber transmits the light impinging on it and the ordered array transmits and replicates the image at the other end. This is shown schematically in Fig. 5.3. An ordered array of optical fibers may therefore serve for the transmission of images. Such image transmission bundles are the basic building blocks of fiberscopes and endoscopes (Epstein, 1982; Wolf, 1979).

The individual fibers in the bundle must be clad fibers. If unclad fibers are used (as in the early endoscopes), light can leak from one fiber into neighboring fibers. This "crosstalk" caused poor quality of the transmitted picture in the early versions of imaging fiberscopes. With the development of clad fibers, it became possible to pack the individual fibers more closely. The ordered optical fiber bundle is similar in many ways to the nonordered ones described above. The fibers

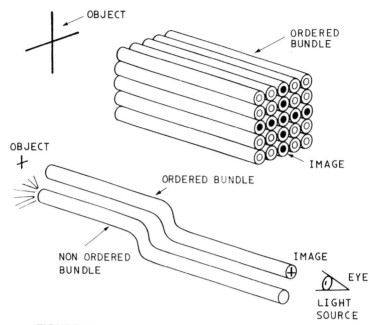

FIGURE 5.3   Picture transmission through an ordered bundle of fibers.

are attached to each other only at their ends so that the whole bundle is reasonably flexible. Ordered bundles may be arranged as Y guides, as shown in Fig. 5.2.

The purpose of an ordered fiberoptic bundle is to transmit the full image of an illuminated object. Optical fibers transmit visible light so well that the image formed by the bundle can be of high quality. Normally the ordered bundle is incorporated into an endoscope and the object is inside the body and illuminated by a nonordered light guide. The imaging bundle must then transmit the image of the object, in full color rendition and with adequate resolution.

5.5 ORDERED FIBEROPTIC BUNDLES—PRINCIPLES

There are limitations to the ability of fiber bundles to transmit images. Some of the limitations are basic; others are related to aspects of the manufacturing processes (Siegmund, 1978; Hopkins, 1976).

5.5.1 Image Transmission—Scientific Limitations

In order to see a detailed image through a bundle, there are two basic requirements: high level of illumination and high resolution.

(i) High illumination: The individual fibers must have a large NA for a high-quality image; therefore the core diameter d must be large and the cladding layer thickness $(D - d)$ preferably small. The large NA and the large core ensure that the maximum amount of light is transmitted from the illuminated object to the eye.

The optical field in the core is not fully confined to the core; some of it leaks onto the cladding layer. If the cladding layer is too thin, some of the light transmitted in one fiber will leak onto the neighboring fibers (crosstalk). This causes deterioration in the quality of the picture. Therefore the cladding layer thickness $D - d$ cannot decrease beyond a certain minimum.

(ii) High resolution: Resolution is the optical property that makes it possible to see small details of the image. The resolution of an optical system can be measured by imaging a line pattern similar to the bar code lines which are found on most items in the supermarket. A series of lines with ever-decreasing width and spacing are used. An optical system of high quality is able to resolve very narrow lines with narrow spacing. In a poor-quality system, the image of the lines will be blurred or will not appear. The resolution is often stated as the number of lines per mm which can be transmitted by the optical system. The resolution depends not only on the optical quality of the system but also on the "contrast" of the individual lines in the image and the illumination of these lines. Under strong sunlight the resolution of the unaided eye is more than 10 lines/mm at standard viewing distance, but under starlight illumination the number may decrease to 2 lines/mm.

It is possible to calculate the resolution capability of an optical fiber bundle (Hopkins, 1976; Morokuma, 1979). The resolution is determined by the diameter d of the cores of the individual fibers in the bundle. The number of lines per mm that can be transmitted by the bundle is limited to approximately $1/2d$. High resolution thus requires small core diameter d.

The following requirements cannot be mutually satisfied: (i) a large core and thin cladding are needed to obtain high illumination; (ii) thicker cladding is necessary to prevent crosstalk; (iii) the size of the core (and the thickness of the cladding) must be reduced in order to obtain high resolution. As a compromise, the core diameter is usually designed to measure approximately $10-20$ μm and the cladding layer $(D - d)/2$ is of the order of $1.5-2.5$ μm.

Under these restrictions, the ratio of the total area of the cores to the whole area of the bundles can be as low as 50%. This places an upper limit on the amount of light that can be transmitted through the bundle, and it also explains why high-quality fibers are needed for imaging. Only with these fibers can a bundle that will give both high spatial resolution and good light transmission be constructed.

One of the problems in transmitting an image of a line object through an ordered bundle is related to the orientation of the line with respect to the bundle. If the object line is parallel to a fiber line in the bundle, the image line will be a fair reproduction of the object. On the other hand, if the object line is in a different direction, as shown in Fig. 5.4a, the image line will look different. Three vertical rows of fibers will transmit the picture and the image will be fuzzy. This is demonstrated in Fig. 5.4b, where the image of the letter N is distorted.

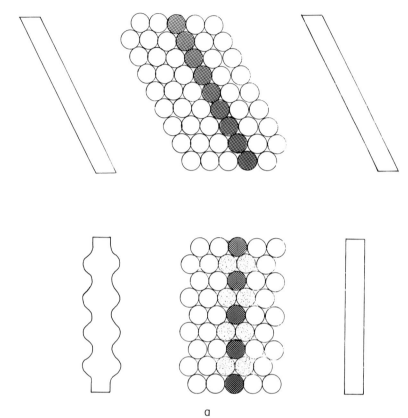

a

FIGURE 5.4   (a and b) Ordered bundles: problems in the transmission of line images. (*Figure continues.*)

5.5.2 Fabrication of Ordered Bundles

An ordered bundle may be fabricated (Epstein, 1982; Siegmund, 1978) by winding a clad fiber on a precision drum. The windings are carefully aligned— unlike the case of a nonordered bundle. After winding one layer onto a drum, other layers are continuously wound, one on top of the other, in an orderly fashion. The whole bundle is cut; the fibers on each end of the cut are glued together, and the ends are polished. Tens of thousands of fibers with diameters of 10–20 μm can be bunched into one bundle.

Alternatively, a small bundle of fibers called a multifiber can be fabricated. An assembly of carefully aligned and ordered lengths of clad fibers is the starting point. The assembly is heated in a furnace until the glass softens and a compound fiber is formed. This compound fiber is treated in the same way as a preform; one end is heated and a multifiber is drawn from this end. The multifiber thus contains many smaller fibers inside. A typical multifiber may contain 100 fibers, each with a diameter of 5 μm. The outer diameter is relatively small, and the multifiber is

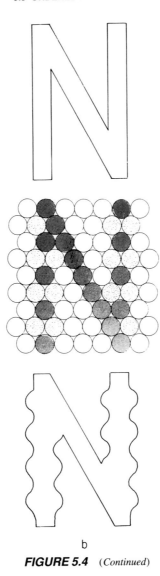

b

FIGURE 5.4 *(Continued)*

thus quite flexible. Several multifibers can be aligned in a bundle. The end result is a full ordered bundle, which can have a high number of fibers (and high resolution). The number of fibers in the multifiber is sometimes large (e.g., 1000). In this case the bundle is not flexible, but rigid, and is called an image conduit. These bundle types are found in thin rigid endoscopes, such as the ones that may be used in dentistry.

Finally, a bundle of fibers may be fabricated by a leaching process, also known as the three-glass (3G) process. Special optical fibers, made of a preform consisting of a core and two cladding layers, are used during this procedure. The inner cladding is a layer of regular glass that has a lower refractive index than the core. The outer cladding layer is made up of a special glass which is soluble in acids. Many multifibers, whose manufacture was described earlier, may now be ordered and heated to form a large, solid multifiber. Each end of the large multifiber is coated with a thick layer of protective plastic material. The whole assembly is then immersed in an acid bath. The acid dissolves the acid-soluble (outer) cladding layer, leaving the inner cladding and the core of the many fibers intact. The individual fibers are thus separated from each other and a flexible fiber results. The protective plastic layer is removed at the two ends and they are cut and polished. Because the acid-soluble glass does not dissolve on the two ends of the bundle, the original ordering of the fibers remains unaltered. The end result is an ordered flexible bundle, known as a leached bundle.

5.5.3 Imaging Bundles—Engineering Limitations

Apart from the limitations on image transmission due to physical optics (as discussed in Section 5.5.1), there are also limitations set by the manufacturing processes.

Stray light: One of the problems with imaging bundles is the effect of unwanted background light—stray light. The stray light is often transmitted through the cladding layers of the individual fibers. This light is transmitted through the bundle and gives an undesirable level of background light that reduces the contrast of the transmitted image.

Defects: Another serious problem with imaging bundles is that of defective fibers in the bundle. During the fabrication process and polishing of the end face, fibers can be damaged or broken. Damaged single fibers that cannot transmit light appear as dark spots. Optical defects and dark spots also appear if a bundle is sharply bent and individual fibers are broken. The presence of these defects constitutes a fixed pattern noise which degrades the quality of the whole picture.

Ordering: In some manufacturing processes, there is a possibility that a few of the fibers will be displaced. The order of the fibers in the input end will not be identical to the order at the output end; the quality of the picture will thus be lowered.

Degradation: Imaging bundle quality may change with time and usage. The individual fibers are not as strong and resistant as the fibers used in communications and are thus more susceptible to mechanical degradation. Imaging bundles which are used in endoscopes are bound to be bent and flexed. This mechanical handling often causes progressive damage and, as a result, endoscopes have a limited lifetime, depending on how often and how carefully they are used.

5.6 FIBERSCOPES AND ENDOSCOPES—FUNDAMENTALS

The fiberscope as an imaging instrument became possible with the perfection of illuminating and imaging optical fiber bundles. A schematic drawing of this

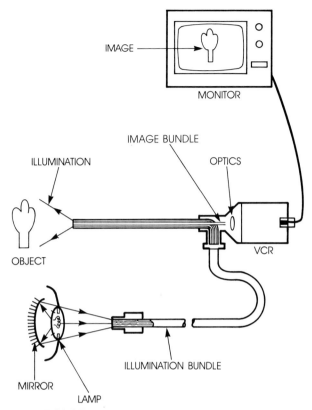

FIGURE 5.5   Fiberscope: longitudinal cross section.

instrument is shown in Fig. 5.5. There is a great advantage in attaching an objective lens to the distal end of the ordered bundle. An image of the object is formed on the distal end face of the ordered bundle and is transmitted through the bundle to the proximal end. Another compound lens system (an eyepiece) is used to facilitate viewing of the image at the proximal end. A photographic camera or a television (video) camera can also be connected to the proximal end of the bundle with a special optical adaptor. The image can then be recorded with film or magnetic tape.

Rigid fiberscopes are based on a rigid image conduit. Flexible fiberscopes, on the other hand, incorporate flexible bundles. One or two light guide bundles are often positioned at the side of the image bundle for illumination. Alternatively, individual fibers at the distal end of the nonordered bundle may be ordered in a circle around the imaging bundle (like the one shown in Fig. 5.1). This saves space and may result in more uniform illumination of the object.

In most ordinary cameras, one can change the distance between the lens and the photographic film, thus making it possible to focus on near or far objects. In principle, one could use the same technique in fiberoptic imaging. By changing the distance between the objective lens and the fiber bundle, one could adjust the focusing of the fiberscope. Adding a distal focusing mechanism complicates the fiberscope; the optical setup is often designed to provide a fixed focus with a large depth of focus (much as in inexpensive cameras).

The fiberoptic *endoscope* is a medical system that makes use of the *fiberscope* for imaging inside the body. In addition, this system contains several ancillary channels that are used for the introduction of thin mechanical tools or for the introduction of liquids inside the body. In this chapter we discuss the optical and mechanical properties of the fiberoptic bundles themselves. The structure and the applications of the medical endoscopes are discussed in Chapter 6.

5.7 FIBEROPTIC IMAGING SYSTEMS—ADVANCES

Advances in fiberoptic imaging may be divided into two groups: (i) new optical methods that make use of imaging fiberoptic bundles, which are discussed in this section, and (ii) applications of fiberoptic endoscopes, which are discussed in Chapter 6.

5.7.1 Thin and Ultrathin Fiberscopes

There is a need to develop fiberscopes whose outer diameter is less than 2 mm. Such fiberscopes will be used for endoscopy in cardiology, pulmonary medicine, neurosurgery, otolaryngology and dentistry. We will arbitrarily distinguish between thin fiberscopes with outer diameters of $1-2$ mm and ultrathin fiberscopes with outer diameters <1 mm. The thin fiberscopes may consist of up to 10,000 individual fibers of typical diameter 5 μm, and the ultrathin ones may consist of $3000-5000$ individual fibers of typical diameter $3-4$ μm. A microphotograph of an ultrathin imaging bundle is shown in Fig. 5.6. It consists of 4000 individual fibers, each of diameter 3.7 μm. The fibers are made of GeO_2-doped silica glass and the outer diameter of the bundle is 0.25 mm.

Let us consider, for example, the angioscope. This endoscope is intended for visualization in the cardiovascular and the peripheral arterial system. The desired outer diameter of such a device has to be less than 2 mm. It should enable physicians to see at least 80% of the coronary arteries, with an optical resolution high enough for a thin suture to be clearly visible (i.e., about 10 lines/mm). In order to achieve such resolution, the number of fibers in the imaging bundle of the fiberscope must be several thousand, and the depth of field needs to be at least 20 mm.

One such fiberscope (D'Amelio *et al.*, 1985) consists of a 1-mm imaging bundle. A beam-shaping light guide consists of a few thicker (50-μm) fibers that are arranged on the circumference of the endoscope around the imaging guide. This arrangement ensures more uniform illumination. Another thin fiberscope (Tsumanuma *et al.*, 1988), whose optical properties have been studied in detail, is based on fused silica. Its high resolution and good mechanical properties point the way toward successful applications of similar endoscopes.

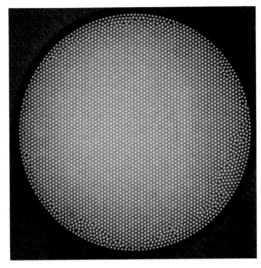

FIGURE 5.6   Microphotograph of the cross section of an optical bundle. (Courtesy of Mitsubishi.)

5.7.2 Magnifying Fiberscopes

In an optical system a viewed object may look larger than its real size; the ratio between the sizes of the object and its image is called magnification. The imaging part of the fiberscope normally consists of a distal objective lens, a fiber bundle, and a proximal ocular lens. The magnification in this case depends on two factors: the magnification by the ocular lens (eyepiece) and the magnification of the objective lens. The latter is determined not only by the lens but also by the distance of the viewed object from the lens. Objects that are closer to the objective lens will be magnified more than objects that are far away (Hopkins, 1976).

Lenses of ordinary endoscopes are fixed and the endoscope is designed to give reasonably good focusing at distances of 5–100 mm. In an attempt to obtain higher magnification, it is not possible to bring the object closer to the endoscope, because the object will be out of focus. In order to reach this goal, different endoscopes have been designed and built. By incorporating a distal focusing mechanism in the endoscope, it possible to bring the object within 2 mm of the objective and obtain magnification higher than $10 \times$. New magnifying endoscopes with magnification higher than $170 \times$ are under development (see Section 6.5.3).

5.7.3 Holographic Endoscopy

Fiberoptic imaging, like most imaging systems, provides a two-dimensional image of an object. Looking through the endoscope, the physician cannot obtain a three-dimensional representation, but rather obtains a "flat" two-dimensional picture. In principle, the fiberscope may consist of two imaging bundles that transmit two images to the two eyes of the observer. A stereo endoscope based on this principle has been demonstrated (Fujimura *et al.*, 1979) but has not yet gained wide acceptance.

In ordinary photography, an object is illuminated by sunlight or by artificial light and a camera is used to project the image of this object onto a photographic film. Processing of the film results in a picture. Illuminating this picture again by sunlight or artificial light produces a two-dimensional representation of the object. A totally different image-recording method, called holography, is based on laser illumination and results in three-dimensional photography. Holography involves a laser beam that is split in two. One part (the illumination beam) is used to illuminate an object, and the reflected light from the object impinges on a special photographic film. The second part of the beam (the reference beam) is sent directly onto the same film. After special processing, the film (now called a hologram) may be viewed under laser beam illumination and a three dimensional picture is observed.

The first attempt to incorporate holographic methods with endoscopy was made almost 20 years ago (Hadbawnik, 1976). A miniature holographic setup was attached to the distal tip of a rigid endoscope and holographic pictures were taken. This method was later adapted to fiberoptic endoscopy. A film was incorporated in a miniature "holocamera" at the distal end of the endoscope. A laser beam sent through the fiberoptic bundle of the endoscope was used as the illumination beam, and the beam reflected from the object impinged on the film. A single mode fiber was inserted through the ancillary channel of the endoscope; a fraction of the laser beam was sent through and also impinged on the film, thus forming a hologram. The film was then removed, processed, and viewed under laser light. A major problem with this method is that special endoscopes need to be developed and that the image is not obtained in "real time."

Alternatively, holograms may be formed when film is placed outside the endoscope. The illumination beam is sent through the imaging bundle of the endoscope to illuminate an object, and the reflected light is transmitted back onto the film. A reference beam is sent directly onto the film, and again the film is processed and viewed under laser light. It is preferable not to use ordinary endoscopes but to use an endoscope in which the individual fibers in the imaging bundle are single-mode fibers. Such bundles have been used experimentally.

Over the last few years, there has been significant progress in holographic endoscopy (Podbielska, 1992; von Bally, 1988). It has been demonstrated that three-dimensional images could be obtained in cavities inside the human body. It has also been shown that these images could in principle have very high resolution.

5.7.4 Image Enhancement

In the field of image processing, a picture is dissected into picture elements that are then stored in a digital form. Sophisticated computer methods may be used to change the contrast in the picture and to enhance desired features. Image processing methods have been proposed in endoscopy but they have not yet been widely utilized.

References

Allan, W. B. (1973). *Fiber Optics: Theory and Practice*. New York: Plenum Publishing.

D'Amelio, F. D., DeLisi, S. T., and Rega, A. (1985). Fiber optic angioscopes. *Opt. Eng.* **24**, 672–675.

Enoch, J. M. (1967). The retina as a fiber optics bundle. In: Kapany, N. S. (Ed.), *Fiber Optics Principles and Applications*, pp. 372–396. New York: Academic Press.

Epstein, M. (1982). Fiber optics in medicine. *Crit. Rev. Biomed. Eng.* **7**, 79–120.

Fujimura, O., Baer, T., and Nimi, S. (1979). A stereo fiberscope with a magnetic interlens bridge for laryngeal observation. *J. Acoust. Soc. Am.* **65**, 478–480.

Hadbawnik, D. (1976). Holographic endoscopy. *Optik* **45**, 21.

Harden, D. (1969). Ancient glass. I. Pre-Roman. *Archeol. J.* **125**, 46–72.

Hecht, E. (1987). *Optics*, 2nd ed. Reading, MA: Addison-Wesley.

Hirschowitz, B. I., Curtiss, L. E., Peters, C. W., and Polland, H. M. (1958). Demonstration of a new gastroscope, the "FIBERSCOPE." *Gastroenterology* **35**, 50–53.

Hopkins, H. H. (1976). Optical properties of the endoscope. In: Berci, G. (Ed.), *Endoscopy*, pp. 3–26. New York: Appleton-Century-Crofts.

Hopkins, H. H. (1976). Physics of the fiberoptic endoscope. In: Berci, G. (Ed.), *Endoscopy*, pp. 27–68. New York: Appleton-Century-Crofts.

Kapany, N. S. (1967). *Fiber Optics Principles and Applications*. New York: Academic Press.

Levi, L. (1980). *Applied Optics*, Vol. 2. New York: Wiley.

Mogi, M., and Yoshimura, K. (1989). Development of super high density packed image guide. *Proc. SPIE* **1067**, 172–181.

Morokuma, T. (1979). Endoscopy. In: Wolf, H. F. (Ed.), *Handbook of Fiber Optics: Theory and Applications*, pp. 429–464. London: Granada.

Podbielska, H. (1992). *Holography, Interferometry and Optical Pattern Recognition in Biomedicine II*. Proc. SPIE, Vol. 1647. Bellingham, WA: SPIE.

Siegmund, W. P. (1978). Fiber optics. In: Driscoll, W. G., and Vaughan, W. (Eds.), *Handbook of Optics*. New York: McGraw-Hill.

Tsumanuma, T., Tanaka, K., Chigira, S., Sanada, K., and Inada, K. (1988). The ultrathin silica based imagefiber for the medical usage. *Proc. SPIE* **906**, 92–96.

von Bally, G. (1988). Holographic endoscopy. *Proc. SPIE* **952**, 2.

Wolf, H. F. (1979). *Handbook of Fiber Optics: Theory and Applications*. London: Granada.

6

Endoscopy

6.1 INTRODUCTION

The previous chapter discussed the optical principles utilized in manufacturing light guides, imaging bundles, fiberscopes, and endoscopes. This chapter offers several examples of actual systems being used in medicine. It is not intended as a guide for endoscopists but rather as an overview. Detailed books and review articles (Berci, 1976; Kawahara and Ichikawa, 1987; Salmon, 1974; Sivak, 1987) are mentioned in the bibliography.

The various uses of fiberoptic endoscopes in different parts of the body are described and discussed. In principle, the various endoscopes are similar to each other, although distinct differences were introduced to make the endoscopes more suited to each particular discipline. Traditionally, the various endoscopes were given different names according to the discipline; most of them are self-explanatory. For example, esophagoscopes are instruments which enable the physician to view the esophagus internally. The first instrument of this kind was a rigid tube, and later ones contained a series of lenses. The fiberoptic endoscope used for viewing the bronchi should be called a bronchofiberscope, but since this book considers only fiberoptic endoscopes, this will be referred to as a bronchoscope. The same applies to all other endoscopes. The following is a partial list of some endoscopes arranged in alphabetic order: angioscopes for veins and arteries, arthroscopes for the joints (an alternative is the orthoscope), bronchoscopes for the bronchial tubes, choledochoscopes for the bile duct, colonoscope for the colon, colposcope for the vagina, cystoscope for the bladder, esophagoscope for the esophagus, gastroscopes for the stomach and intestines, laparoscope for the peritoneum, laryngoscope for the larynx, ventriculoscope for the ventricles in the brain.

6.2 ENDOSCOPIC IMAGING SYSTEMS—FUNDAMENTALS

The endoscopic system consists of several subsystems: the endoscope itself; subsystems that provide illumination, irrigation, and suction; and auxiliary medical subsystems such as electrocautery and biopsy forceps (Kawahara and Ichikawa, 1987). The endoscope will first be described and then the various subsystems.

6.2.1 The Endoscope

In Section 5.2, the principles related to the operation of the fiberscope and the fiberoptic endoscope were described. This section details the major components of a medical endoscope. The simplest example is the gastroscope. A schematic drawing of the instrument is shown in Fig. 6.1. The endoscope consists of three major parts: (i) the insertion tube section, with the distal end; (ii) the handpiece (control box) section, with the control knobs and other functions; and (iii) the connecting cable section, with the light guide and tubes connecting air and water supplies.

6.2.1.1 The Distal End

The distal end of the endoscope houses the distal tips of several of the optical and mechanical components. These are the tip of one or two light guides, which provide illumination; the distal tip of the imaging bundle; and a miniature objective lens (or a complex lens system), which is attached to the tip of this bundle. A thin, flat protective window that is easy to clean may cover the tip to protect it from contamination by blood or debris. The distal end of the endoscope is often immersed in fluids, secretions, or blood, which tend to hinder viewing. A flushing

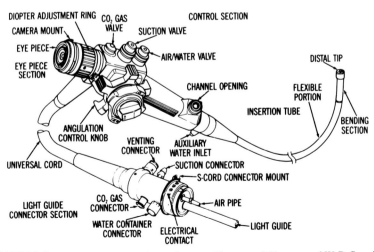

FIGURE 6.1 Schematic drawing of an endoscope. (Courtesy of Olympus and W. B. Saunders.)

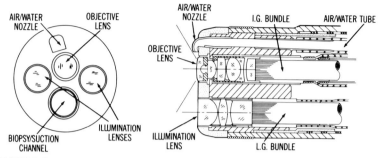

FIGURE 6.2   Distal end: (*left*) front view and (*right*) cross section. (Courtesy of Olympus and W. B. Saunders.)

port is supplied through which a stream of water can be sprayed on the distal tip of the imaging system to keep it clean. The objective lens system varies from endoscope to endoscope. Different lenses give different fields of view and the focal distance may be different. Apart from a difference in lenses, the optical systems differ in other respects. Some endoscopes receive a picture in a forward direction. These are called forward-looking endoscopes. Side-viewing endoscopes are sometimes needed, for example, when viewing the biliary tree from inside the duodenum. By attaching a miniature prism to the objective lens, the sidewalls can be seen in greater detail; the light exit, the lens-washing exit, and the protective windows are all situated on the side at the distal end. Some endoscope designs are borrowed from industry, where the prism is rotated (by a remote control) inside the distal end. This facilitates extended viewing on all sides of the endoscope. The distal end may also contain one or several outlets that are used for irrigation, aspiration, suction, or insertion of thin instruments. In some cases, the instrument outlet (also called the operating port) is protected by a bridge that is raised when the instrument is introduced through the endoscope. Two cross sections of a typical distal end are shown in the Fig. 6.2.

6.2.1.2   Flexible Shaft

The flexible shaft holds the light guide, imaging bundle, and ancillary tubes together. The shaft is connected at one side to the distal end and at the other side to the proximal end of the endoscope. This shaft is often constructed of steel mesh and can be strengthened further by a steel spiral. This construction serves two purposes. It stiffens the endoscope, giving it mechanical support, and protects the delicate optical and mechanical components inside from being crushed or kinked. It also enables the physician to control the whole endoscope mechanically. The physician can rotate or torque the proximal end of the shaft and the distal end will follow suit. Some endoscopes have a more flexible section of the shaft, next to the distal end, which can be bent by a remote mechanical control. Usually there are thin wires that run inside the shaft, connected to rings in the more flexible section. By using the thumbs, the physician can rotate the knobs that pull these wires and

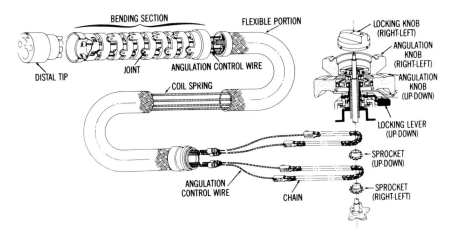

FIGURE 6.3 Angulation mechanism. (Courtesy of Olympus and W. B. Saunders.)

thereby move the distal end in two perpendicular directions. The flexible shaft is sheathed by a plastic jacket that is biologically inert and that forms a hermetic seal. It protects the inner parts of the endoscope from interaction with water, blood, or other biological fluids. The flexing and angulation mechanisms are shown in Fig. 6.3.

6.2.1.3 Proximal End

The proximal end of the endoscope includes viewing optics, controls, and several ports. These are often housed in a handpiece held by the physician. All the control knobs, buttons, levers, and the photographic equipment are connected to this handpiece, as was shown in Fig. 6.1. Each component is discussed separately.

(i) Optics: The optical system of the endoscope may have a fixed-focus arrangement. All objects within a certain distance from the distal tip will be seen clearly (for example, all objects within the range 5 mm to infinity). Other endoscopes have a focusing unit in addition to the eyepiece. Photographic equipment may be attached to the eyepiece, using various adapters, in order to facilitate recording a whole sequence of events. A teaching attachment is available that makes it possible for two people to look through the endoscope; this has now largely been superseded by video facilities.

(ii) Controls: As mentioned above, the distal end is connected by wires to knobs or levers that are located on the proximal handpiece. In some endoscopes, the physician can also control the flow of air or water to the distal end.

(iii) Ports: Several ports serve different purposes. Some of the ports are for passage of liquids, such as saline solution or drugs, others are for pressurized gases, such as air or CO_2. There are ports for the aspiration of gas or the suction of liquids. Surgical instruments and auxiliary optical fibers can be introduced through the operating port.

6.2.1.4 Cables

An assortment of cables connect the endoscope to the supply system. The proximal end of the light guide is connected to the light source. Air insufflation and aspiration tubes are connected to an air pump, and the liquid ports are connected to liquid reservoirs or to liquid suction subsystems. In normal endoscopic systems, the whole assembly of cables is incorporated in one umbilical cord that is sheathed in a plastic jacket.

6.2.2 Supply Subsystem

This subsystem supplies illumination and irrigation and it often provides compressed gases or pumps gases out through the endoscope.

6.2.2.1 Light Sources

Appropriate illumination is essential for photography and optimal vision. The light source must have high intensity and a desired color distribution. To obtain the desirable spectral characteristics, the light source must be operated at high temperatures. At lower temperatures, the light is not natural (i.e., too red) and the color rendition is poor. A high light output is often obtained by passing high currents through the lamp. However, this results in a relatively short lamp life, as well as undesirable heating effects in the lamp housing. The following modifications overcome some of these problems:

(i) Quartz halogen lamp: In ordinary tungsten lamps, the filament evaporates easily. A thin deposit of tungsten forms on the inside of the quartz or glass envelope, preventing light from emerging. With the introduction of a small amount of halogen gas (e.g., iodine) into the lamp, this process is reduced. Iodine serves in a regenerative cycle to keep the bulb free of tungsten, resulting in longer lamp life and a higher light output than with ordinary lamps. In some quartz halogen lamps, there is a small reflector that helps to focus the light from the lamp onto the input face of the light guide. The reflector may even be a selective reflector that focuses only visible light but not IR. This minimizes the amount of unwanted IR light, thus reducing the heating at the input end of the light guide.

(ii) Mercury and xenon arc lamps: These lamps are small but powerful. The gases in these lamps are highly pressurized and, when operated at a high current, they emit relatively high intensity in the visible part of the spectrum. The major problem in arc lamps is that they cannot be operated directly from the 110-V supply and require a special power supply.

6.2.2.2 Pumps

The supply system contains an air pump that provides pressurized air for the insufflation tube. The pressurized air can drive water from a reservoir through a thin tube to the distal end. Water is often used for cleaning the lens or for other purposes, such as cooling power fibers. The same air pump or a different one can be used for pumping out gases or liquids. The pumped-out gases are released into the atmosphere, whereas the liquids are pumped out into a reservoir. The ability to pump out gases and debris is vital in endoscopic laser surgery.

6.2.3 Auxiliary Mechanical Devices and Subsystems

Many surgical instruments can be made small enough to be inserted through the ancillary port(s) of an endoscope, especially in large endoscopes such as the gastroscope or the colonoscope. Some of the more common instruments are mentioned here and a few are shown schematically in Fig. 6.4.

(i) Grasping forceps: Grasping forceps or jaws are attached to the tip of a thin and flexible metal tube. Metal wires run inside this tube, and by pulling them the physician can open the jaws. The thin tube is inserted into the operating port and pushed through to the distal end. If necessary, the physician can push the grasping jaws outside the distal end and open them. This can be used to grasp a foreign object that has to be removed from the body (see Fig. 6.4a,b).

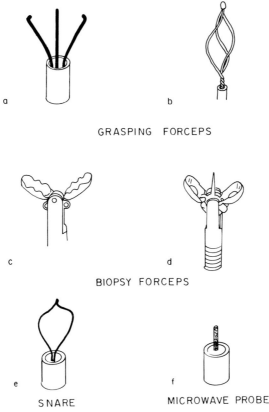

a   b

GRASPING   FORCEPS

c   d

BIOPSY   FORCEPS

e   f

SNARE   MICROWAVE   PROBE

FIGURE 6.4   Endocopic accessories: forceps, snare, and microwave probe.

(ii) Biopsy forceps: These are similar to the grasping forceps except that the jaws are shaped like small cups with sharp edges. When these jaws are closed, they can remove a small piece of tissue for biopsy. There is sometimes a central spike at the center of the device between the jaws that serves to keep the tissue stationary during the biopsy procedure (see Fig. 6.4c,d).

(iii) Cytology brushes: These devices consist of a flexible metal wire (or a spring) that ends in a metal brush at its distal tip. The brush is sometimes protected by a thin Teflon tube. The device is again inserted through the operating port and the distal tip is driven out of the endoscope. The brush is then pushed, under endoscopic viewing, out of the Teflon tube. It may be used to collect cells from a lesion. Once the brush is withdrawn from the tube, the device can be pulled out with the sample cells intact for analysis.

(iv) Snare: This device is built from a thin wire which enters through the operating port; it continues all the way to the distal end and returns to the operating port. If the whole wire is pushed forward, a loop forms, as shown in Fig. 6.4e. The loop can be snared around a piece of tissue, such as a polyp. If a high-frequency current is passed through the wire, the heat generated by the current may cut the tissue and coagulate the blood at the same time, preventing bleeding from the tissue.

(v) Cutting tools: Tiny cutting tools such as surgical knives and scissors can also be attached to the end of flexible wires and inserted through the ancillary channels of an endoscope. Miniature drill bits have been introduced through thin endoscopes and operated by rotating the flexible wire at high speed.

6.2.4 Photographic Subsystem

Two types of photographic equipment can be attached to the eyepiece:

(i) Still camera: A regular camera is attached to the eyepiece by a simple adaptor. Normally a "reflex" (through the lens) camera is used with electronically controlled flash.

(ii) TV/video: A TV (video) camera is attached to the eyepiece, making it possible to record an entire therapeutic procedure. Modern video cameras make use of a new imaging system that is based on a miniature imaging device called a charge-coupled device (CCD), which is discussed in Section 6.4. This is an added convenience for endoscopy.

6.3 ENDOSCOPIC IMAGING SYSTEMS—PRINCIPLES

A large number of commercially available endoscopic systems are used by physicians. The basic principles of all the instruments are the same. It is worthwhile to describe two simple systems: the traditional endoscope and the thin or ultrathin endoscopes, which are promising but still in their embryonic stage. A full endoscopic system is shown schematically in Fig. 6.5.

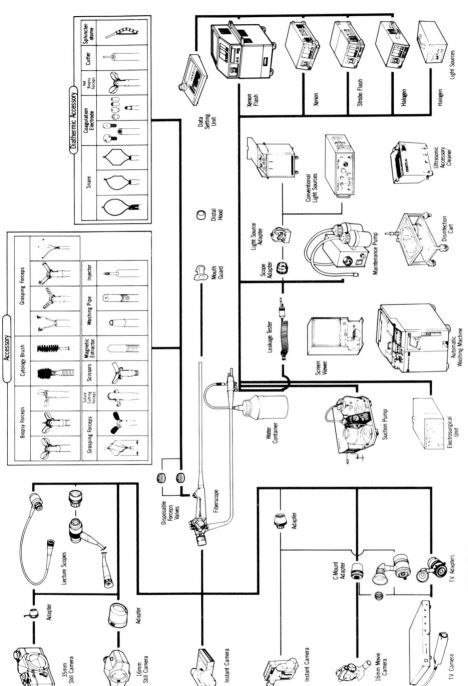

FIGURE 6.5 Schematic drawing of a commercial endoscopic system. (Courtesy of Olympus and W. B. Saunders.)

TABLE 6.1 Data on a Few Endoscopes

Endoscope	Length (mm)	Outer diameter (mm)	Channel diameter (mm)	FOV (deg)	Depth of focus (mm)	Bending (deg)
Colonoscope	1700	11–13	3	120	5–100	230
Gastroscope	1000	9–11	2–3	100	3–100	240
Bronchoscope	500–700	5–6	2	100	3–50	160
Angioscope	1000	1–2	<1	80	3–20	180

6.3.1 The Common Endoscope

A few characteristic parameters of some of the important types of endoscopes are given in Table 6.1.

Endoscopes, like other optical or nonoptical systems, are designed and built to perform specific goals. Some of the features that can be specified are as follows:

(i) Transmission range: e.g., 0.4–0.9 μm. This is the range of wavelengths which is well transmitted by the individual fibers. It determines the color rendition.

(ii) Resolution: e.g., 3–5 lines/mm. This number determines the ability of the optical system to discriminate details in the image. The higher the number, the better the ability to see small details in the image.

(iii) Magnification: e.g., 1–10× (depending on the observation distance). This is the ratio between the size of the image observed through the endoscope and the object size.

(iv) Depth of focus: e.g., 2–50 mm. The distance over which a clear image is obtained.

(v) Field of view (FOV): e.g., 10°–50°. The angular field which can be imaged through the endoscope.

(vi) Minimum bending radius: e.g., several millimeters. The minimum radius over which the endoscope can be bent without danger of damage due to breaking of optical fibers.

(vii) Bending angle: e.g., +90°–90°: The maximum angle over which the distal tip may be bent. In normal endoscope the tip may be bent in two directions. There are endoscopes in which the tip may be bent in one direction at an angle larger than 90°.

Figure 6.6a illustrates a modern endoscope and Fig. 6.6b and c show pictures taken by such instruments.

6.3.2 Thin Endoscopes—Angioscopes

Since the turn of the century, there has been great interest in direct visualization of the interior of the heart chambers and blood vessels. In the early 1920s

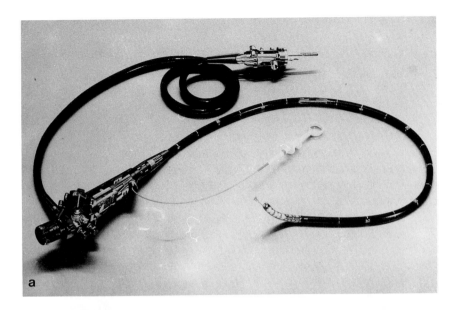

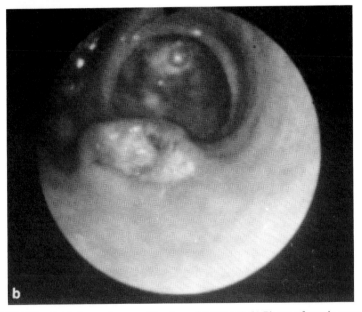

FIGURE 6.6 (a) Modern endoscope. (Courtesy of Olympus.) (b) Picture of gastric cancer taken by a gastroscope. (Courtesy of Olympus.) (c) Picture of grasping of a stone taken by a choledocoscope. (Courtesy of Olympus.) (*Figure continues.*)

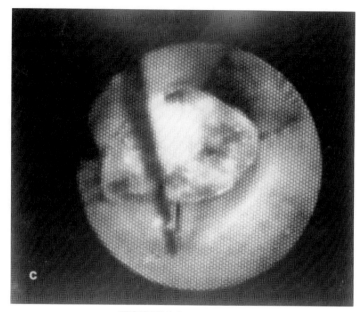

FIGURE 6.6 (*Continued*)

(Allen and Graham, 1922) a rigid endoscope (i.e., angioscope) was used to observe the heart chambers of a dog. In the late 1930s and the 1940s, rigid endoscopes were used to visualize cardiac chambers in humans during surgery. In these investigations, blood was displaced in order to facilitate visualization. This was achieved by first using a jet of clear fluid, such as saline. Later, balloon-tipped angioscopes were used and imaging was carried out through the clear balloon during open chest surgery (thoracotomy). All these endoscopes were fairly thick and not very useful. In the 1980s this changed with the development of thin fiberoptic endoscopes that were used for percutaneous endoscopy of the pulmonary and coronary arteries.

We will distinguish between thin endoscopes, whose outer diameter is 1–3 mm, and ultrathin endoscopes, whose outer diameter is less than 1 mm. In this section we discuss only the former. In Section 5.7.1 we discussed some of the optical properties of fiberscopes that serve as the basis for the fabrication of thin and ultrathin endoscopes. These are far more difficult to manufacture than ordinary-size endoscopes and only the thin endoscopes contain an ancillary channel. One such thin endoscope, the angioscope, which is designed to obtain an image from the inside of the arteries, will serve as an example (Trauthen, 1990) for this family of endoscopes.

(i) Optical image: The image quality of the endoscope is limited by the number of optical fibers in the imaging bundle. In a typical coronary artery, the maximum lumen diameter is 3 mm, limiting the outer diameter of an angioscope to under 3 mm. In order to obtain useful resolution, the imaging bundle of the thin endoscope has to have at least 10,000 fibers.

(ii) Mechanical properties: The individual fibers, which are thin and fragile, make the thin endoscopes susceptible to damage. After prolonged use, the fibers can break, leading to the appearance of dark spots in the image. This is one of the most severe problems of thin endoscopes, and endoscope manufacturers are trying to improve the mechanical quality. Another approach is to have good-quality disposable angioscopes.

(iii) Angulation: In a thick endoscope, the physician can apply a torque to its proximal end. By rotating the whole flexible shaft, the physician can bring the imaging bundle to any desired rotational position. This is not yet possible with most thin endoscopes; friction between the endoscope and the blood vessel wall makes it difficult to transmit rotational movement along the shaft. Although stiffening elements such as spiral wires or braids can be added, they require an unacceptable increase in the diameter of the endoscope. A metal guide wire is sometimes inserted, allowing the thin endoscopes to slide over the wire to the desired location. Alternatively, a guide catheter is introduced and the thin endoscope is inserted through this catheter.

(iv) Image enhancement: In angioscopy, it is difficult to distinguish between plaque and the blood vessel wall. The fluorescent imaging methods (Section 6.5), however, may help to clarify this and make the necessary distinctions (Andersson et al., 1987; Andersson-Engels et al., 1991). UV light can be used to advantage when the natural luminescence of the plaque is different from that of the blood vessel wall.

(v) Sterilization: It is not easy to circulate liquids or gases through the ancillary channel of thin endoscopes. It is therefore difficult to sterilize these endoscopes. This problem has not been solved satisfactorily. One of the solutions is to use disposable thin endoscopes. Several companies have already begun to develop such products.

In the future, it will be possible to insert thin angioscopes into a blood vessel and perform diagnosis and therapy. There has been a rapid improvement in the performance of the thin endoscopes (Mizuno et al., 1989), and some of the latest clinical results are discussed in Chapter 9.

6.4 ENDOSCOPIC IMAGING—ADVANCES

6.4.1 Videoendoscopy

In the early days of fiberoptic endoscopy, the optical quality of the imaging bundles was not satisfactory. One solution involved attaching a miniature camera to the distal end of the endoscope. The imaging bundle was used only to locate a desired area. Illumination was provided by a light guide and still photographs were taken with the camera. This has changed with the development of higher-resolution imaging bundles.

During the last few years, there has been tremendous progress in the development of miniature imaging devices that are based on semiconductor silicon technology. One of the most advanced imaging devices is based on charge-coupled devices (CCDs). This device, manufactured by microfabrication techniques, is an array of tens or hundreds of thousands of light-sensitive elements, in addition to the necessary electronic circuits needed to operate these elements, all on a small Si chip whose area is less than 1 cm^2. The whole device is a miniature TV camera, with its full circuitry and controls.

One of the problems in fiberoptic endoscopy is resolution, or the ability to discriminate between neighboring elements. Video endoscopes have been made with CCD devices placed at the distal tips, replacing the imaging bundles, while the nonordered fiberoptic bundle served for illumination. When an image is projected on the CCD, the circuits translate it to digital electronic signals. These are sent via thin wires through the endoscope to a video monitor, where they can be stored on a videotape. The image may then be viewed on a TV screen. This is rather similar to the way an ordinary video camera is used to take pictures that are viewed with a videocassette recorder (VCR). The important difference is the size of the camera. Typically, the size of the endoscopic CCD device is 10–15 mm on a side and it may contain over 1000 × 1000 elements; such cameras have been used in gastroscopy (Petrini, 1987; Sivak, 1992).

Videoendoscopes may be compared to ordinary fiberscopes with respect to resolution, color performance, and viewing angle. Several studies have compared these parameters qualitatively (Knyrim *et al.*, 1990; Satava *et al.*, 1988). The resolution of an optical fiberscope depends on various parameters, including the resolution of the optical bundle, which depends on the size of the individual optical fiber that transmits each element of the picture. An image bundle with a diameter of several millimeters may consist of up to 40,000 fibers, each with a diameter of about 7 μm. Similarly, the resolution of the CCD chip of a fixed area depends on the individual light-sensitive element which "captures" one picture element, called a pixel. The size of a pixel may be less than 4 μm, and CCDs containing 4 million have been reported. In tests that were performed with various videoendoscopes it was found that the typical resolution is roughly 100 μm, which is comparable to that of a fiberscope.

The CCD chip without modifications cannot distinguish between colors. Obtaining a color picture requires the use of complementary techniques such as illuminating the object with consecutive pulses of red, blue, and green or depositing red, blue, or green optical filters on the individual light-sensitive elements. The consecutive pulses reduce the speed at which an image is obtained; this might introduce blurring. The use of three light-sensitive elements in each picture element reduces the resolution. Such techniques for obtaining color pictures are widely used in ordinary videophotography and have also been used in video endoscopy. Experiments show that the color rendition of videoendoscopes is good. The videoendoscope and a picture taken with it are shown in Fig. 6.7a and b.

At present, videoendoscopes have demonstrated that they can generate pic-

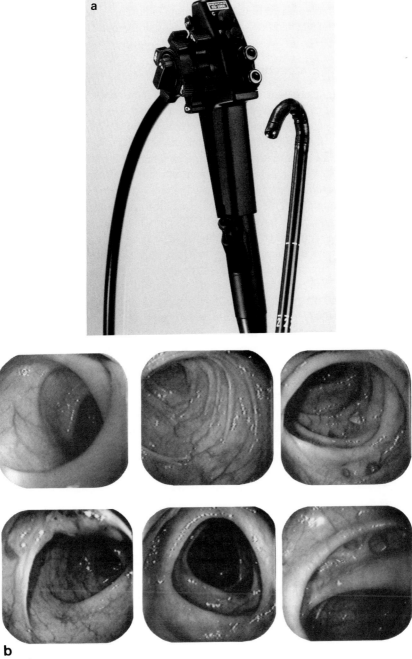

FIGURE 6.7   (a) Videoendoscope. (Courtesy of Pentax.) (b) Pictures taken through a videoendoscope. (Courtesy of Pentax.)

tures comparable to those obtained with regular fiberscopes. The CCDs are still too large to be incorporated in thin or ultrathin endoscopes, and the systems are still more complicated and less reliable than regular fiberoptic ones.

6.4.2 Ultrathin Endoscopes

In some applications it is desirable to have endoscopes whose diameter is smaller than that of thin endoscopes. Ultrathin endoscopes have been developed with outer diameters of less than 1 mm. The imaging bundle of such an endoscope consists of less than 3000 fibers each of diameter of roughly 2–5 μm (see Section 5.7.1). The outer diameter of the imaging bundle is often less than 0.3 mm and the optical resolution is lower than that of a thin endoscope. This endoscope is too thin to include a practical ancillary channel. Also, it is not yet possible to incorporate an angulation mechanism in such a thin endoscope.

Ultrathin endoscopes have been used clinically for imaging of peripheral and coronary arteries. Preliminary experiments have also been conducted in other medical disciplines. In otolaryngology they have been used for imaging in various parts of the respiratory system. They were used inside the salivary tunnel, the paranasal sinus, and even the bronchus of a 2-year-old child. In urology they have been used inside the ureter and the renal pelvis. In orthopedics they have been used for imaging inside the spinal cord and in small joints in the upper and lower limbs. Such endoscopes will also be useful in gynecology, ophthalmology, dentistry, and neurosurgery.

As mentioned in Section 6.3.2, one of the problems of thin endoscopes is that of sterilization of the ancillary channel. An interesting possibility is to introduce the ultrathin fiberscope into the body through a thin guide catheter. This catheter will have an outer diameter of less than 3 mm and an ancillary channel of diameter roughly 1 mm. This catheter will preferably have an angulation mechanism which will be used to guide it to its proper position. Once in place, the ultrathin fiberscope will be inserted in the ancillary channel of the catheter. It is expected that the catheter will be disposable but the ultrathin endoscope could be used many times.

Thin and ultrathin endoscopes are shown in Fig. 6.8.

6.5 ENDOSCOPIC DIAGNOSTICS—ADVANCES

The imaging capabilities of the endoscope serve, largely, for diagnostic purposes. By inspection, the physician can usually tell if the organ is healthy or not; sometimes, however, the image does not supply enough information. Several methods can be used to improve the imaging capabilities and the endoscope offers several other options which help in diagnosis. In this section, only a few of these are discussed. Other examples are discussed in later chapters dealing with fiberoptic biosensors.

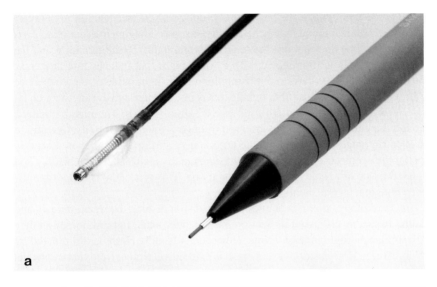

a

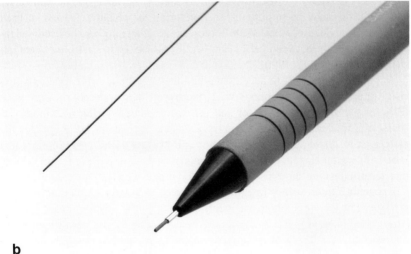

b

FIGURE 6.8 Photographs of (a) a thin endoscope and (b) an ultrathin endoscope. (Courtesy of Mitsubishi.)

6.5.1 Fluorescence Endoscopy

In Section 3.4.2 we discussed two types of tissue fluorescence: the autofluorescence from tissue itself (also called endogenous) and the fluorescence of agents added to the tissue (also called exogenous). These two types of laser-induced fluorescence (LIF) may be used for diagnosis at a given area on the tissue. Typically, light at a given wavelength is sent through an optical fiber and excites a characteristic tissue luminescence that is transmitted through an optical fiber for optical processing. This is the basis for fiberoptic sensors (see Section 7.1). Alternatively, the same method may be used in endoscopic imaging to accentuate parts of the image. One does not use the regular "white light" for illumination, but the tissue under study is illuminated with a wavelength that induces luminescence. The excitation may be carried out using the regular light guide of the endoscope or through a special optical fiber bundle that is inserted through the ancillary channel. UV or blue lasers are often used as excitation sources. The image obtained through the endoscope may be observed through a filter which transmits only in the spectral range of the peak of the emission spectrum. This filter is placed in front of the eyepiece and enables the physician to distinguish between healthy and diseased tissue or to study the state of the tissue. This procedure is called fluorescence imaging, as was mentioned earlier.

In cardiology, there is a problem in distinguishing between plaque and healthy tissue, as discussed in Section 3.5.2. Under white light illumination they look similar through an angioscope. However, if healthy tissue is illuminated by UV light (e.g., HeCd laser), the emission is markedly different from that of plaque. Fluorescent imaging, based on autofluorescence, (Andersson *et al.*, 1987; Hoyt *et al.*, 1988) is possibly a useful way to distinguish between plaque and blood vessel wall; this may be important for monitoring the ablation process in laser angioplasty (see Section 9.2.5). Autofluorescent imaging may also be used in dentistry to distinguish between caries and healthy tissue, or in gastrointestinal endoscopy (Cothren *et al.*, 1990).

An alternative way is to accentuate the image by the use of exogenous (extrinsic) fluorescence. In Section 3.5.2.1, the use of hematoporphyrin derivative (HPD) for the diagnosis of cancer and for fluorescence imaging of tumors was discussed. The characteristic (exogenous) red fluorescence of HPD enables the endoscopist to distinguish between tumor and healthy tissue in the endoscopic image (Andersson-Engels *et al.*, 1991; Baumgartner *et al.*, 1987; Profio, 1988). Other sensitizing dyes have also been tried for similar applications.

Exogenous fluorescence imaging may also be used in examining noncancerous tissue. For example, photosensitizers are retained by atherosclerotic plaque, as mentioned in Section 3.5.2.2. It should be noted that the autofluorescence imaging worked well during *in vitro* experiments. In animal and preclinical experiments, however, there was significant interference due to the autofluorescence of blood and the results were inferior. It therefore seems reasonable to assume that exogenous fluorescence imaging will be better for clinical applications.

6.5.2 Staining

The nature of the observed tissue is not always clear in endoscopy. In gastroscopy, for example, the fine details of the gastrointestinal mucosa cannot be observed in some cases. Staining methods may be used to facilitate better imaging. The most common method is to introduce a thin catheter through the ancillary channel of an ordinary endoscope. A proper coloring agent (Ida and Tada, 1987) is then injected into the catheter. Three different coloring schemes have been tested clinically:

(i) Contrast: The coloring agent is accumulated in grooves or irregularities in the tissue and thus increases the contrast of the image. In gastroenterology, indigo carmine has been used for better observation of small lesions.

(ii) Staining: There is a difference in the uptake of some coloring agents in healthy and diseased tissue. In cardiology, for example, a coloring agent such as β-carotene may be preferentially absorbed in atherosclerotic plaque. In gastroenterology, methylene blue has been used for staining of the gastric mucosa.

(iii) Reaction: The coloring agent reacts with specific cells, which may thus be identified. In gastroenterology, Congo red dye has been used to define the acid-secreting mucosa. Different dyes have been tried for the identification of cancer cells.

6.5.3 Magnification

Magnifying rigid gastroscopes were used in the early 1950s to observe faveoli (diminutive pits) in the gastric mucosa. In the 1960s and 1970s magnifying fiberscopes were developed, first with magnification of $15 \times$ and later with magnification of $30 \times$. These gastroscopes made it possible to observe the mucosal surface in patients and distinguish between healthy tissue and peptic ulcers or gastric cancer (Nishizawa *et al.*, 1980; Takemoto and Sakaki, 1987). They may be used for pathophysiologic studies of the gastric mucosa and other tissues. This demonstrates the potential use of such fiberscopes as diagnostic tools.

6.5.4 3D Imaging and Size Determination

In regular endoscopy there are a few fundamental limitations that affect diagnosis:

(i) Three-dimensional (3D) imaging: The image obtained is two-dimensional, which sometimes makes diagnosis difficult. One of the proposed methods for overcoming this difficulty (see Section 5.7.3) is endoscopic holography. Another method is stereoendoscopy based on a special endoscope which contains two imaging bundles and two eyepieces—one for each eye.

(ii) Size of objects: The endoscopic image does not provide quantitative information about the size of objects. It is not possible to tell the difference between large objects that are far away and smaller objects that are near, although experienced endoscopists can make a reasonably good estimate. There have been proposals to project a laser light beam through the endoscope and generate a grid of known size on an object (Yamaguchi *et al.*, 1988). Holographic interferometry may also make it possible to perform quantitative measurements endoscopically. The methods mentioned in this section have not yet been widely used.

6.6 ENDOSCOPIC THERAPY—FUNDAMENTALS

In numerous cases the endoscope can be used for therapy as well as diagnostics. In this section, a few of these cases are discussed in general. More specific examples are given in Chapter 9. Therapeutic modalities involving lasers are not discussed at this stage.

6.6.1 Surgery

Operative instruments such as cutting knives or biopsy forceps can be introduced through the ancillary channel (see Section 6.2.3). In this category, one may also include rotating cutting tools or abrasive tips which remove tissue. In cardiological and vascular surgery, several tools have been tested for atherosclerotic plaque removal (atherectomy) inside blocked arteries (see Section 9.2). The same technique can be utilized to remove cartilage and bone, located in joints, via an arthroscope.

6.6.2 Electrosurgery and Coagulation

One of the problems associated with regular cutting tools is how to stop the initiation of bleeding. This problem can be reduced if the cutting is done by an electrode, such as a thin wire heated by an electrical current. The hot wire can cut soft tissue at the point of contact and simultaneously coagulate the blood from small vessels (Odell, 1987). An endoscopic electrosurgery system is shown schematically in Fig. 6.9.

In order to perform useful cutting with a heated wire, it is necessary to heat the wire with an electrical frequency higher than 10^5 Hz (e.g., radio frequency). At a lower frequency, muscles or nerves are stimulated, causing contractions. An electrode in the shape of a thin wire can be inserted through the ancillary port. One side of the wire is connected to a high-frequency power supply and the other

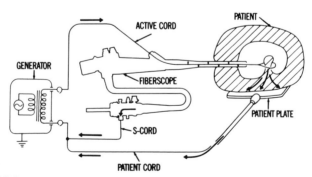

FIGURE 6.9 Endoscopic electrosurgical system. (Courtesy of Olympus and W. B. Saunders.)

to the body of the patient. The electric loop is closed by connecting the body of the patient to the power supply. If a high enough current flows through the wire, its temperature will be sufficiently high to cut the tissue on contact. The cutting can be accompanied by bleeding. If this occurs, the current passing though the wire can be reduced. Now the heat is not sufficient to cut but is sufficient for hemostasis.

6.6.3 Microwave Cutting and Coagulation

Microwave radiation at 2450 MHz is well absorbed in water and heats it efficiently. This is the basic principle for heating food in domestic microwave ovens. Biological tissue contains water and can therefore be heated by this means.

Electromagnetic radiation at high frequencies can be transmitted through a coaxial cable that consists of an inner copper wire, an insulating (plastic) layer, and an outer copper braid. This type of cable is used for transmitting TV signals in cable TV or in video systems. The same cable can also be used to transmit microwave energy. At the end of the cable, a small segment of the inner core that protrudes out of the insulation (see Fig. 6.4f) may serve as an antenna. The proximal end of the cable is connected to a suitable power supply. Microwave power of tens of watts can be transmitted through the cable and be radiated out through the antenna. The cable, which may be fairly thin (e.g., 1 mm) can be inserted into the ancillary channel of an endoscope (Tabuse *et al.*, 1985). Microwave radiation sent through the cable can reach the distal end of the endoscope. The radiation will be most efficiently coupled with tissue if the tiny antenna is inserted a few millimeters into the tissue. The microwave radiation causes heating and can be used for blood coagulation (e.g., hemostatic treatment of bleeding ulcers) or for localized hyperthermia. If the intensity of the radiation is high, it may also be used for cutting. Microwave resection is accompanied by coagulation and is potentially useful even for organs which are rich in blood vessels, such as the liver.

A new application of such microwave devices is in urology for treatment of the prostate gland. The device is inserted through a catheter, and the heat generated by the microwave shrinks the prostate gland. This is a minimally invasive treatment that may be an alternative to transurethral resection or open prostatectomy (surgical removal of the gland).

6.6.4 Grasping

Section 6.2.3 mentioned that various grasping tools can be inserted into the ancillary channel. Grasping forceps may be used to grasp a foreign body and hold it while the endoscope is being pulled out, removing the foreign body (see Fig. 6.4a). This may not sound like a surgical procedure, but it replaces one. The same is true when a physician excises the gallbladder or a large tumor which would otherwise have been removed by open surgery.

6.6.5 Sample Acquisition—Biopsy

Sometimes it is necessary to perform diagnostic measurements on some tissue samples. In the past, this was impossible without invasive surgery. The endoscope has been widely used to obtain tissue samples without open surgery. The procedures to be described involve minimally invasive procedures. An example of tissue removal is the cytology brush mentioned in Section 6.2.3, with which the physician collects cell samples which are sent to the laboratory for analyses. A second example is biopsy using the biopsy forceps, also mentioned in Section 6.2.3. The physician can remove a small sample of tissue under direct vision, and the sample can then be analyzed.

6.7 ENDOSCOPIC ULTRASOUND IMAGING—PRINCIPLES

A fiberoptic endoscope may be inserted into a blood vessel or cavity in the body, providing an image of the inner surface of the cavity or the blood vessel. If it is inserted into the gastrointestinal tract, the physician can evaluate the mucosal surface and perhaps detect early cancer. The depth and the extent of the cancer are not obvious from the optical image obtained. A supplementary ultrasound technique called ultrasonography (or echography) may help in this case (Strohm and Classen, 1987).

Sound waves are waves which propagate in elastic media at frequencies between 20 and 20,000 Hz. Their speed of propagation in a material depends on the elasticity and the density of the material. Waves of a similar nature but much higher frequency (1–10 MHz) are called ultrasonic waves. Ultrasonic waves that are transmitted into a piece of matter will be reflected from various discontinuities in the sample, including the edges, foreign material, or changes in physical density. The "echo" signals that are obtained may be used to measure the thickness of the sample, to analyze its inner structure, or to detect the presence of defects or discontinuities. This is the basis of ultrasonography or ultrasound imaging.

Ultrasound may be generated and detected by devices whose size is of the order of 1 mm. Such a device may be attached to the tip of a catheter or an endoscope and pointed perpendicular to the long axis of the catheter. Ultrasound signals are sent and the echo signal is used to study tissue structure. As early as 1956, such a catheter was used for intracardiac investigations in dogs (Cieszynski, cited by Bom and Roelandt, 1989). During the past 30 years, these methods have been improved and opened up a new area of ultrasound imaging. One important development is rotation of the single device so that a two-dimensional image of the surroundings of the catheter is obtained. The spatial resolution of the image is better than 1 mm. Another important development is the construction of an array of devices that operate in sequence (a phased array) and generates a two-dimensional picture. This is shown schematically in Fig. 6.10.

The ultrasound imaging system may be miniaturized so that its outer diameter

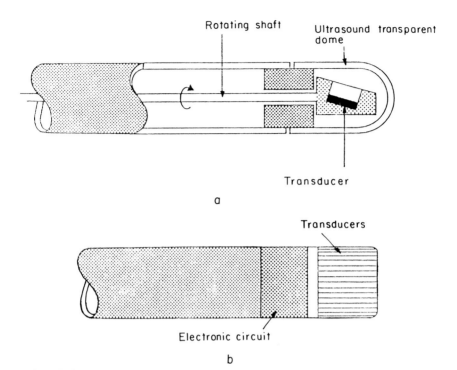

FIGURE 6.10 Endoscopic ultrasound imaging system: (a) single rotating transducer and (b) phased array of transducers.

is of the order of 1–2 mm. A photograph of the tip of an early prototype of such a device used for intravascular imaging is shown in Fig. 6.11. A tiny ultrasonographic system may be attached to the distal tip of an ordinary fiberoptic endoscope. The endoscope tip is guided to a desired location, for example to the upper gastrointestinal tract. High-resolution ultrasonographic investigation of the gastrointestinal tract and its surrounding organs may then be performed. Endosonographic studies have been performed by such dedicated ultrasound endoscopes in urology, gynecology, and gastroenterology. The development of miniature ultrasound devices opened up the possibility of using ultrasonography even in cardiology, in conjunction with angiography (Bom and Roelandt, 1989). The ultrasound system is inserted into arteries and the two-dimensional image obtained supplies information about the thickness of blood vessel walls and the presence (and composition) of atherosclerotic plaque.

There are other options for using ultrasound. The ultrasound probes may be designed to be so thin that they would be inserted through the ancillary channel of an ordinary fiberoptic endoscope (Kimmey *et al.*, 1990). This probe may help in guiding the endoscope tip to a lesion of interest. Another interesting option is the TULIP (transurethal laser-induced prostatectomy). In this method a probe which

FIGURE 6.11   A miniature ultrasound catheter. A photo of a match is shown for comparison. (Courtesy Dr. E. Gussenhoven.)

consists of an ultrasound transducer and an Nd:YAG power fiber is introduced into the prostatic ureter through a cystoscope. The prostate tissue is destroyed by laser coagulation and the process is controlled via ultrasound imaging. The clinical applications of ultrasound imaging are discussed further in Chapter 9.

References

Allen, D. S., and Graham, G. (1922). Intracardiac surgery. A new method. *JAMA* **79**, 1028–1030.

Andersson, P. S., Montan, S., Persson, T., Svanberg, S., Tapper, S., and Karlsson, S. E. (1987). Fluorescence endoscopy instrumentation for improved tissue characterization. *Med. Phys.* **14**, 632–636.

Andersson-Engels, S., Johansson, J., Svanberg, K., and Svanberg, S. (1991). Fluorescence imaging and point measurements of tissue: applications to the demarcation of malignant tumors and atherosclerotic lesions from normal tissue. *Photochem. Photobiol.* **53**, 807–814.

Baumgartner, R., Fisslinger, H., Jocham, D., Lenz, H., Ruprecht, L., Stepp, H., and Unsold, E. (1987). Fluorescence imaging device for endoscopic detection of early stage cancer—industrial and experimental studies. *Photochem. Photobiol.* **46**(5), 759–763.

Berci, G. (1976). *Endoscopy*. New York: Appleton-Century-Crofts.

Bom, N., and Roelandt, J. (1989). *Intravascular Ultrasound*, Dordrecht: Kluwer Academic Publishers.

Cothren, R. M., Richards-Kortum, R., Sivak, M. V., Fizmaurice, M., Rava, R. P., Boyce, G. A., Doxtader, M., Blackman, R., Hayes, G. B., Feld, M. S., and Petras, R. E. (1990). Gastrointestinal tissue diagnosis by laser induced fluorescence spectoscopy at endoscopy. *Gastroentintest. Endosc.* **36**, 105.

Hoyt, C. C., Richards-Kortum, R. R., Costello, B., Sacks, B. A., Kittrell, C., Ratliff, N. B., Kramer, J. R., and Feld, M. S. (1988). Remote biomedical spectroscopic imaging of human artery wall. *Laser Surg. Med.* **8**, 1–9.

Ida, K., and Tada, M. (1987). Chromoscopy. In: Sivak, M. V. (Ed.), *Gastroenterologic Endoscopy*, pp. 203–217. Philadelphia: W. B. Saunders.

Kawahara, I., and Ichikawa, H. (1987). Fiberoptic instrument technology. In: Sivak, M. V. (Ed.), *Gastrointestinal Endoscopy*, pp. 20–41. Philadelphia: W. B. Saunders.

Kimmey, M. B., Martin, R. W., and Silverstein, F. E. (1990). Endoscopic ultrasound probes. *Gastrointest. Endosc.* **36**, S40–S46.

Knyrim, K., Seidlitz, H., Vakil, N., and Classen, M. (1990). Perspective in "electronic endoscopy." *Endoscopy* **22**, 2–8.

Mizuno, K., Arai, T., Satomura, K., Shibuya, T., Arakawa, K., Okamoto, Y., Miyamoto, A., Kurita, A., Kikuchi, M., Nakamura, H., Utsumi, A., and Takeuchi, K. (1989). New percutaneous transluminal coronary angioscope. *J. Am. Coll. Cardiol.* **13**, 363–368.

Nishizawa, M., Kariya, A., Kobayashi, S., and Shirakabe, H. (1980). Clinical application of an improved magnifying fiber clonoscope with special reference to the remission features of ulcerative colitis. *Endoscopy* **12**, 76–80.

Odell, R. C. (1987). Principles of electrosurgery. In: Sivak, M. V. (Ed.), *Gastroenterologic Endoscopy*, pp. 128–142. Philadelphia: W. B. Saunders.

Petrini, J. L. (1987). Video endoscopy. In: Sivak, M. V. (Ed.), *Gastroenterologic Endoscopy*, pp. 253–271. Philadelphia: W. B. Saunders.

Profio, A. E. (1988). Review of fluorescence diagnosis using porphyrins. *Proc. SPIE* **907**, 150–156.

Salmon, P. R. (1974). *Fibre Optic Endoscopy*. New York: Grune & Stratton.

Satava, R. M., Poe, W., and Joyce, G. (1988). Current generation video endoscopes—a critical review. *Am. Surg.* **54**, 73–77.

Sivak, M. V. (1987). *Gastroenterologic Endoscopy*. London: W. B. Saunders.

Sivak, M. V. (1992). Electronic endoscopy. *Endoscopy* **24**, 154–158.

Strohm, W. D., and Classen, M. (1987). Endoscopic ultrasonography. In: Sivak, M. V. (Ed.), *Gastroenterologic Endoscopy*, pp. 182–202. Philadelphia: W. B. Saunders.

Tabuse, K., Katsumi, M., Nagai, Y., Kobayashi, Y., Noguchi, H., Egawa, H., Aoyama, O., Mori, K., Yamaue, H., Azuma, Y., and Tsuzi, T. (1985). Microwave tissue coagulation applied clinically in endoscopic surgery. *Endoscopy* **17**, 139–144.

Takemoto, T. and Sakaki, N. (1987). High magnification endoscopy. In: Sivak, M. V. (Ed.), *Gastroenterologic Endoscopy*, pp. 220–230. Philadelphia: W. B. Saunders.

Trauthen, B. (1990). Technical considerations for angioscopic imaging. *Proc. SPIE* **1201**, 580–583.

Yamaguchi, M., Okazaki, Y., Yanai, H., and Takemoto, T. (1988). Three dimensional determination of gastric ulcer size with laser endoscopy. *Endoscopy* **20**, 263–266.

7

Fiberoptic Diagnosis

7.1 INTRODUCTION

There is a pressing need for improvements in diagnostic medical techniques. Present techniques can be divided into two classes: invasive and noninvasive. The noninvasive methods, such as x-rays, computed tomography (CT), magnetic resonance imaging (MRI), or ultrasound imaging, do not cause pain to the patient and are potentially safe. The invasive methods, which require penetration into the body, are still necessary because they are often more accurate and reliable, providing information that cannot yet be obtained otherwise. Optical fibers are the basis for a relatively new family of "least invasive" diagnostic tools called optical fiber sensors.

A sensor is a device in which changes in input energy (called a measurand) produce corresponding changes in another form of energy (called a signal). For example, in a thermocouple, changes in temperature (measurand) produce changes in voltage (signal). The signal may be used to monitor and to control the measurand. In the case of a thermocouple, the signal is used to control heating or cooling systems. In this example, the measurand is thermal. Other measurands could be mechanical (e.g., pressure, flow, weight), radiant (e.g., light intensity or color), or chemical (e.g., chemical composition, pH, Pco_2). A plethora of sensors have been used in science and industry, and many of them have been used in medicine. One of the newer sensor technologies is the fiberoptic sensor, which has been tried for some industrial applications (Culshaw, 1982). In these sensors, changes in the measurand cause changes in the light transmitted in an optical fiber. It is expected that similar sensors will be widely used in medicine.

The requirements of a good sensor system are:

(i) Specificity: The measurement of only one parameter at a time without being affected by other parameters.
(ii) Sensitivity: The ability to measure small changes in a given parameter.
(iii) Accuracy: The ability to give an accurate and stable measurement over a long period.
(iv) Cost: A low cost would enable the sensor to be disposable.

In principle, fiberoptic biomedical sensors have all these attributes, including the following advantages:

(i) Miniaturization: They are based on very thin fibers and can easily be inserted into various places in the body through a catheter or hypodermic needle. This enables the physician to perform measurements *in vivo* and in real time.
(ii) Biocompatibility: All the materials from which the fibers and transducers are made are nontoxic and compatible with biological tissue.
(iii) Accuracy: Fiberoptic sensors may be in principle fast, stable, and accurate.
(iv) Safety: Since no electrical wires and circuits are introduced into the body, these are safe sensors.

At present, blood samples are generally sent to a laboratory for a chemical analysis of one or more of their constituents. The laboratory may be far from both the patient and the physician. There are frequent delays and unavoidable errors in the chemical analysis; thus, therapeutic decisions may not be based on complete and accurate information. An alternative is the use of miniaturized sensors that can be inserted into the body of the patient. Theoretically, it is possible to make a sensor small enough to fit into blood vessels without interfering too much with the blood flow. The sensor should be able to perform chemical analysis in real time and should be disposable to avoid hepatitis, AIDS, and other transmissible diseases. In the past, it was believed that miniature electronic devices such as integrated circuits (commonly known as chips) could perform this task. Various electronic circuits can measure pressure or temperature, and if the components are exposed to the blood, they may respond to chemical changes such as pH. Many of these devices have been successfully tried in the laboratory, or *in vitro*. However, tremendous difficulties are involved in using these techniques *in vivo*. Simultaneously, with the development of better optical fibers, there was accelerated development of new fiberoptic sensors for industrial and military applications. In a typical sensor, a thin optical fiber is part of a complex optical system. The fiber is placed in a location and is used to measure some parameter such as temperature, pressure, or magnetic field strength. Many of these techniques have been adopted for medical applications.

In principle, a thin optical fiber may be inserted into the blood to measure blood pressure, pH, or partial pressure of oxygen. In practice, there are many fiberoptic sensors which have been suggested and several which have been tried *in vivo*. A few have already been developed for commercial products and are being offered for clinical trials. It is a fairly new technology, however, with hurdles still to be crossed. This chapter describes some of the important fiberoptic sensors and their principles of operation. Detailed descriptions of these sensors are given in review articles (Hall, 1988; Milanovich *et al.*, 1984; Peterson and Vurek, 1984; Seitz, 1984; Vurek, 1984; Walt 1992; Wolfbeis, 1987) and books (Wise and Winegard, 1991; Wolfbeis, 1992).

It is again fascinating to observe that nature has discovered the benefits of optical fiber sensors. An interesting use of optical fibers in nature is in plants (Mandoli and Briggs, 1985). One example is related to seedlings that are in darkness under the soil, such as oat seeds. A shoot emerging from the seedling behaves like an ordered bundle of fibers. These fibers transmit light by total internal reflection over a distance of a few centimeters and apparently the bundle may also transmit a crude image. The optical fibers facilitate monitoring of light intensity, the spectral distribution of light, and the direction and duration of light above the ground. Some of these sensing functions may take place inside the light-guiding tissues themselves. The plants therefore make use of true fiberoptic sensors that are probably responsible for the coordination of plant physiology.

Some insects use fiberoptic sensors as well (Callahan, 1977). It is well established that the male moth is attracted to the female through pheromones—acetate compounds emitted by the female and considered as the female "sex scent." There are indications that this acetate has a characteristic luminescence under visible light excitation, with narrow emission lines in the far IR ($\lambda > 10$ μm). The male moth has short antennal sensilla (spines) which act as dielectric waveguides that are transparent in the IR. The male moth detects the characteristic emission, and is attracted to the female, through the use of these IR fiberoptic sensors.

7.2 FIBEROPTIC MEDICAL DIAGNOSIS—FUNDAMENTALS

In Section 3.5, the fundamentals of laser-assisted diagnostics were described. It was shown that blood, urine, or tissue can be illuminated by a laser beam, and by analyzing the reflected (or the luminescent) light, information is obtained about that biological fluid or the tissue. This method is clearly suitable for use with fiberoptic systems. By inserting two thin optical fibers through a catheter into the blood vessel and then transmitting light at a particular color through one of the fibers, the reflected light from the blood is collected and transmitted through the second fiber to a detector. The signal obtained from the detector is proportional to the reflection by blood of this particular color. If this process is repeated for a number of colors, the reflection spectrum of blood is obtained. Such a spectrum can reveal the oxygen saturation of blood. In this case, the ends of the fibers are bare and diagnosis is made by direct interaction between light and blood; the sensors are called *direct* fiberoptic sensors. Other cases involve an optical transducer called an *optode* that is attached to the end of one or two optical fibers. Light sent through the first fiber interacts with this optical transducer, which in turn interacts with the body fluids. The light sent back through the second optical fiber again carries information about the body fluids. These are called *indirect* fiberoptic sensors. In both cases, the optical fibers are just the vehicles for transmitting the light into the body and back again. The major advantage of optical fibers is that they enable measurements to be made inside the human body—even inside the heart or other major organs.

The various analytical methods used in science and medicine can be broadly divided into physical and chemical methods. In a physics laboratory, instruments such as gauges or thermometers are used to measure physical parameters such as pressure, temperature, or flow. Fiberoptic sensors that measure tissue temperature, blood pressure, or blood flow are called *physical sensors*. In a chemistry laboratory, reagents and chemical reactions are used to measure pH or contents of chemicals in a sample. An optical fiber sensor that measures pH, Po_2, Pco_2, or glucose content in the blood can be called a *chemical sensor*. In this category are sensors that make use of sophisticated biochemical techniques for measuring parameters of the blood and concentration of drugs such as penicillin. These are called *biochemical sensors*.

It should be emphasized that this is a relatively young field of research. The novel fiberoptic sensors still suffer from problems such as slow response, accuracy, and reliability. Nevertheless, several companies already manufacture and distribute fiberoptic sensors of the types mentioned. Details of some of these sensors are given in the following sections.

7.3 FIBEROPTIC BIOMEDICAL SENSORS—PRINCIPLES

All fiberoptic sensors transmit one beam of light into the body through an optical fiber. The beam interacts directly or indirectly (through a transducer) with biological fluids or with biological tissue. As a result of the interaction, another beam of light emerges that is transmitted back to the outside world through an optical fiber. The comparison of the input beam to the output beam provides information about the biological sample. This diagnostic method allows the optical interrogation of tissue situated deep inside the body without invasive surgery. A general fiberoptic sensor system (either physical or chemical) used for medical diagnostics is shown schematically in Fig. 7.1. Both direct and indirect sensors are shown.

The input and output beams may be separated by several methods, two of which are shown in the figure. In one method, the input light is focused onto the proximal end of the fiber through a hole in a spherical mirror. The output light which emerges from the proximal end is reflected by the same mirror onto the analyzing optical system. A second method separates input and output beams by a special filter called dichroic mirror. Such a filter can transmit the input beam

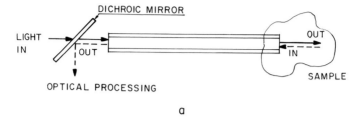

a

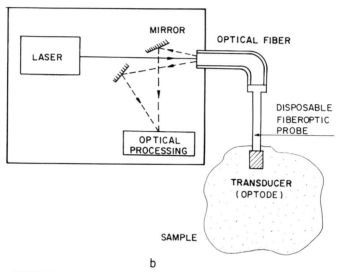

b

FIGURE 7.1 Fiberoptic sensing system: (a) direct and (b) indirect.

from left to right with little attenuation. The output beam (at a different wavelength) that cannot be transmitted back through the filter from right to left is reflected, as shown. Instead of using just one fiber, two fibers may be used: one for the incident beam and one for the reflected (or emitted) light.

7.4 DIRECT FIBEROPTIC SENSORS—PRINCIPLES

Direct fiberoptic sensors are shown schematically in Fig. 7.2. With these sensors, light (i.e., laser light) is coupled into the proximal end of an optical fiber. It is then transmitted to the distal end, which is located in a sampling region. The light emerging from the fiber is an input beam, which interacts directly with the sample, giving rise to an output beam, usually a reflected beam or luminescence. Part of the output beam may be collected by a second fiber (as shown in Fig. 7.2a) and transmitted back to the proximal end. It is then coupled into an optical processing system. The analysis of the output light is used for blood measurements. Alternatively, a nonordered fiber bundle may be used. Half of the fibers are used

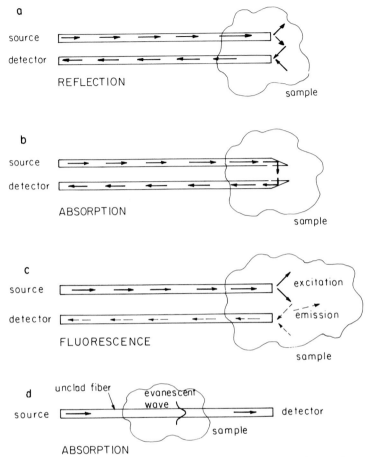

FIGURE 7.2 Direct fiberoptic measurements: (a) reflection; (b) absorption; (c) fluorescence; (d) evanescent wave.

to send the input beam in, and the other half are used to transmit the output beam out; this is called a bifurcated bundle (see Fig. 5.1).

Absorption measurements of tissue can also be performed using two fibers. Light is sent via one fiber and transmitted through a thin layer of tissue, and the transmitted light is returned via a second fiber. The ratio of the intensities of the input to the output light is measured at various wavelengths to give the absorption spectrum of the tissue. This is shown in Fig. 7.2b. The fluorescence spectrum can be directly measured in a similar way, as shown in Fig. 7.2c. The optical absorption in a sample can also be directly measured by so-called evanescent wave absorption (Fig. 7.2d), which is discussed in Section 7.6.

As mentioned, direct sensors have no attached transducers (optodes). The direct sensors can be divided into the two arbitrary categories: physical and chemical sensors.

7.4.1 Direct Physical Sensors

In conventional diagnostics, several optical methods are useful:

(i) *Photometry* (or colorimetry): Monochromatic light is directed onto tissue or blood and the absorption or reflection at this wavelength is measured. If this is performed at many wavelengths, the spectral absorption or the spectral reflection of the sample is obtained. Such spectra are "fingerprints" of specific compounds in tissue (or in blood).

(ii) *Fluorometric method*: Excitation of luminescence in the sample. By measuring the excitation spectrum or emission spectrum of the sample, important information is provided.

Both of these optical methods are direct methods that require no transducer and can readily be applied through optical fibers, as discussed below.

7.4.1.1 Reflection and Absorption

It was mentioned earlier in this chapter and in Section 3.5.1 that reflection or absorption spectroscopy serves as a powerful diagnostic tool when applied to blood or tissue. Optical fiber systems have been developed to carry out measurements inside the body. In this section, the use of such systems for noninvasive measurements (e.g., on the skin) of physical quantities is addressed. Such measurements can, in principle, be performed without optical fibers, but the fibers facilitate access.

Muhl (1986), for example, measured the reflectivity of the skin at 700 nm. It was found that light of this wavelength penetrates deeply (millimeters) into the skin and the reflectivity is strongly affected by the microcirculation of blood. Reflectivity measurements (which are noninvasive) can thus be used to assess the microcirculation and measure blood flow. These measurements have been carried out in areas where blood vessels are close to the surface, such as inside the lips or in the eyes. The transmission of light through some parts of the body, such as the ear lobe, is determined mostly by the blood content of the tissue. Absorption measurements can therefore be easily performed by fiberoptic systems. In all these measurements, the distal end of the fiber can be shaped in different ways, as shown in Fig. 7.3.

7.4.1.2 Temperature Measurement—Radiometry

Several methods which are used for measuring temperature rely on indirect fiberoptic sensors (see Section 7.5.1.2). In this section, a direct method that makes use of a "bare" infrared-transmitting fiber (see Section 4.7.3) is described. Objects at temperatures higher than absolute zero emit radiation known as blackbody radiation. Emission increases as the temperature increases. The formula for the total emission from a blackbody at temperature T is

$$I = \varepsilon\sigma T^4$$

where ε is the emissivity and σ is the Stefan–Boltzmann constant. Bodies at room temperature emit mainly in the far infrared. As the temperature rises, more and

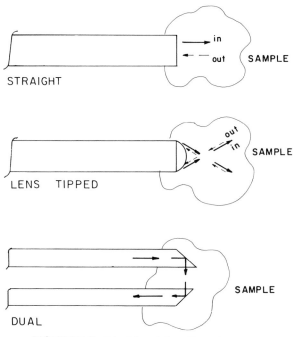

FIGURE 7.3 Distal tips of direct fiberoptic sensors.

more emission appears in the near infrared and finally in the visible. For biological tissue, $\varepsilon = 1$; the total emission is dependent only on the temperature T. If the IR emission is measured, the temperature of the tissue can be determined.

Over the past four decades, the art and science of measuring infrared radiation has been developed by the military for tasks such as detecting vehicles, personnel, and ships in total darkness. A thermographic picture can reveal warm bodies (such as tanks) via their IR emission. The IR emission from each part of an object is measured and the temperature can be displayed. The motor of the vehicle will be warmer than the chassis. Using the same technique, a thermographic system can monitor the surface temperature of the human body. In pathological conditions, such as breast cancer, the affected breast area will be slightly warmer than the healthy breast. This is revealed by a thermal camera which displays the temperature distribution of the chest on a TV monitor. This method is used for detecting the disease. Although IR imaging is less reliable than x-ray mammography for breast screening, it is totally passive and the patients need not be exposed to ionizing radiation.

Radiometer systems or thermographic systems used for measuring temperature by way of IR emission depend on a line of sight between the warm surface and the IR detector. Sometimes no such line of sight exists. It may, however, still be possible to transmit the IR emission from a warm body to a detector via an infrared-transmitting fiber. This method is called IR fiber radiometry.

The fiberoptic radiometry system consists of an infrared detector and a flexible IR fiber. The IR emission from a surface is transmitted through the fiber and is measured by the detector. Preliminary experiments have found that the fiberoptic radiometer is capable of measuring temperature with a resolution greater than 0.1°C.

The feasibility of using the fiberoptic radiometer for microwave heating has been investigated (Katzir et al., 1989) and other fiberoptic techniques have been suggested for hyperthermia and hypothermia (see Section 7.5.1.2). IR fibers may also be used for measuring the temperature of a surface which is exposed to laser radiation during medical laser treatment. It is important to limit the surface temperature to prevent thermal damage to the tissue. This temperature cannot easily be measured by conventional devices such as thermistors or thermocouples.

With fiberoptic radiometry the fibers can be introduced through an endoscope, in order to monitor and control tissue temperature during endoscopic laser surgery or laser welding.

7.4.2 Direct Chemical Sensors

In this section, two direct methods for measuring chemical parameters using fibers are described (Martin et al., 1987). The first measures optical properties such as reflection or absorption in tissue. The second involves fluorescence, either from a natural biological tissue or from some reagents which are introduced into the tissue or blood. Both methods were also discussed in Section 3.5.

7.4.2.1 Fiberoptic Reflection and Absorption Measurements

Oximetry

Oxygen in the blood is carried by the hemoglobin in red blood cells. Oxygen saturation is the percentage ratio between the oxygen content in a given blood sample and the maximum oxygen-carrying capacity. The blood in the arteries should be more than 95% saturated and that in veins about 75% saturated. Measurement of oxygen saturation is useful in cardiology, in monitoring patients during anaesthesia, and in intensive care units. Samples of blood taken from the right side of the heart that show an unusually high saturation ratio may indicate a congenital abnormality of the heart. Blood measurements that show low saturation may be due to reduced capacity of the blood to carry oxygen, low cardiac output, or reduced ability of the cardiac pulmonary system. Optical methods can be used to monitor the oxygen saturation, based on different optical properties of saturated and desaturated hemoglobin. The optical absorption spectrum and the optical reflection spectrum are different for each. Figure 7.4 presents a reflection spectrum of oxyhemoglobin (HbO_2) and reduced hemoglobin (Hb). Two characteristics are obvious:

(i) There is a great difference in the red part of the spectra, in particular at $\lambda = 620$ nm.

(ii) There is no difference at a wavelength of $\lambda = 805$ nm, which is called the isobestic point.

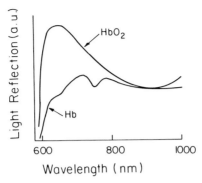

FIGURE 7.4 Determination of O_2 content in blood by photometry.

Polanyi and Hehir (1962) were the first to use fiberoptic reflection spectroscopy for measuring oxygen saturation. They sent an input beam through one bundle of fibers and measured the output beam through another bundle. The ratio $I(660)/I(805)$ was used to determine the oxygen saturation *in vivo*. Fiberoptic oximetry measurements were also carried out by Kapany and Silbertrust (1964).

Volz and Christensen (1979) used a similar technique to determine the oxygen saturation in neonates (newborn). They used two light-emitting diodes (LEDs) as light sources and sent two different wavelengths through two fibers. The reflected light was measured using other fibers. These thin fibers were inserted into a body of a dog through a thin catheter. The oxygen content in the blood differed with the various respiratory gas mixtures. The measured oxygen saturation was compared with the result from a standard blood analyzer; there was good agreement between the results. This system was intended for Po_2 monitoring in neonates, where the fiberoptic probe was inserted into the body through an umbilical artery catheter. It shows the feasibility of using the reflection method *in vivo*. A similar system is now widely used in monitoring of patients by means of a simple finger probe held on with a spring.

7.4.2.2 Fiberoptic Laser-Induced Fluorescence

Fluorescence techniques are widely used as analytical tools, as mentioned in Section 3.5. Laser-induced fluorescence (LIF) techniques are well suited for systems that are based on lasers and optical fibers. This section discusses nonimaging applications in which both the exciting laser beam and the emitted beam are transmitted through single optical fibers or through light guides. These are actually "point" measurements. LIF techniques can also be performed through imaging bundles, which result in fluorescence imaging of an area (see also Section 6.5.1).

Natural Fluorescence (Autofluorescence)

The excitation and emission spectra of many types of tissue have been investigated extensively. By studying the complete spectrum of a tissue, it is often possible to obtain information about the pathology of the tissue (Andersson-Engels *et al.*, 1989, 1990). A few examples of such measurements in medical disciplines are given:

Cancer: The autofluorescence emitted by healthy tissue is different from that of cancerous tissue in both the spectral emission and the temporal behavior. Such differences may be used to detect cancer endoscopically.

Cardiology: In order to deal with plaque blockages inside the arteries, the plaque must be identified during endoscopic imaging. The intrinsic autofluorescence spectrum of plaque has been compared to that of the normal blood vessel wall.

Dentistry: Alfano and Alfano (1985) found that the luminescent emission from a healthy tooth is different from that emitted by a tooth with caries. When excited at 480 nm, the carious areas absorb more than noncarious areas and also emit more red light. Analysis of the emitted light signal can thus reveal whether the tooth is healthy or not. Similar fiberoptic methods have been applied inside the tooth to determine the state of the root canals (Pini *et al.*, 1989).

Fluorescence of Reagents and Dyes in Tissue (Exogenous Fluorescence)

In Section 3.5.2.1 it was mentioned that hematoporphyrin derivative (HPD), which is selectively accumulated in malignant tissue, has a characteristic luminescence at 630 nm that may be used to reveal its presence. HPD fluorescence, an example of a laser-induced fluorescence spectroscopic technique, is a powerful diagnostic tool. Similar methods are easily adaptable both for nonimaging endoscopic diagnosis and for fluorescence endoscopy.

Fluorescence of Reagents or Dyes in Blood—Flow Measurements

Thermodilution, one of the simplest methods for measuring blood flow, entails injecting a small quantity of cool fluid into the heart. Small thermometers (thermocouple or thermistors) are introduced via catheters into several places in the circulation. By measuring the temperature change along the path of the blood, the volumetric blood flow rate can be calculated with sufficient accuracy. Thermodilution is now routinely used clinically (see also Section 7.5.1.3).

An equivalent optical method is based on introducing a luminescent dye into the heart and measuring the rate at which it becomes diluted in the circulation. Using x-ray or endoscopic guidance, an optical fiber can be inserted into the body and guided into an artery. Excitation light sent through the fiber causes luminescent emission from the dye. This luminescence is transmitted back to a detector through the same fiber or through a second fiber. Following the dye injection, the luminescence signal decays with time due to the dilution of the dye. The decay kinetics can be used to calculate the volumetric blood rate flow. Additional information can be obtained from other fiber probes placed in several locations throughout the circulation.

In their early experiments, Polanyi and Hehir (1962) used a dye excited by light at 805 nm. At this wavelength (i.e., the isobestic point) the optical absorption does not depend on the oxygen content (see Fig. 7.4). Therefore, while exciting luminescence, changes in oxygen saturation do not contribute an optical signal and do not interfere with the blood flow results. The same method was used by Voltz and Christensen (1979). More recent experiments (Korb *et al.*, 1989) made use of a dye, indocyanine green, to measure the myocardial blood flow. Experiments were performed percutaneously on mongrel dogs without opening their chests. Using a regular catheter, the dye was introduced into the right atrium of the heart. Two optical fibers were introduced into the circulation, one into the aortic root and one into the coronary sinus. The luminescence signals obtained from these fibers were measured continuously after the injection of the dye. The blood flow was determined from the time dependence of the luminescence. The coronary blood flow in the dogs was varied by the intravenous infusion of various drugs. In each case, the blood flow was measured. These experiments showed the feasibility of using the dye dilution fiberoptic technique clinically.

NADH Fluorimetry

Fluorescence measurements may supply information about alterations in cell metabolism. Such knowledge is valuable in cardiology for monitoring changes in cardiac metabolism and for detecting heart disease. For this purpose, two compounds are studied: nicotinamide adenine dinucleotide (NAD) and reduced nicotinamide adenine dinucleotide (NADH). In normal functioning living cells which use oxygen, the ratio between NAD and NADH varies over a large range. Under extreme conditions, however, such as lack of oxygen (anoxia) or lack of blood supply (ischemia), there is 100% reduction of NAD to NADH. The cell may survive short exposure to these extreme conditions, but prolonged exposure will cause cell death. Clinically, if 100% reduction of NADH can be detected at an early stage, the physician may have a chance to restore the cell viability before irreversible changes take place.

It was found that NADH absorbs UV light at wavelengths in the spectral range 250–400 nm; it then emits visible fluorescence which peaks in the spectral region 460–480 nm. In contrast, NAD does not fluoresce. The luminescence intensity is linearly proportional to the number of NADH excited molecules; using fluorescence measurements, one may therefore determine the intracellular concentration of NADH. This measurement, carried out through optical fibers (Mayevsky and Chance, 1982), is called laser fiberoptic fluorimetry. A practical system (Renault *et al.*, 1985; Renault, 1987) is based on the excitation of fluorescence by a nitrogen laser at 337 nm through an optical fiber. The fluorescence emission, at 480 nm, is detected through the same fiber and a narrow-bandpass filter. The distal tip of the fiber may be inserted into the body through a catheter or an endoscope; the measurement has been performed inside the heart or other organs.

Using NADH laser fluorimetry, cardiologists monitored dead heart tissue (myocardial necrosis) after a heart attack (myocardial infarction) to determine the extent of damage. Laser–fiberoptic fluorimetry may become a powerful diagnostic tool which supplies vital information to surgeons in real time during vascular surgery, "bypass" heart surgery, or neurosurgery. *In situ* laser fluorimetry which provides continuous measurements may have significant advantages over the more traditional biochemical analysis of blood specimens.

Plasma Emission from Stones

When high-energy pulses impinge on urinary or biliary stones which are immersed in liquids, they generate a "physical" plasma (see Section 3.4.5) and the resulting shock wave fragments the stones. This procedure in urology is performed using a laser catheter and is the basis of laser lithotripsy (see Section 9.11.4). The laser-produced plasma is hot (10^5 °C) and is accompanied by light emission (Teng *et al.*, 1987). During the first 0.5 μsec, the emission is very broad and structureless. Later, after the plasma cools, sharp emission lines appear. This characteristic emission, generated by ions (e.g., Ca^+) or neutral atoms (e.g., Ca), may be transmitted back through the power fiber and used for diagnostic or control purposes. For example, it may ensure that the distal tip of the fiber is aimed at a stone and not at normal tissue.

A similar plasma emission appears during the ablation of calcified plaque by high-energy laser pulses. This can be used, in principle, to monitor and control plaque removal in laser angioplasty.

7.5 INDIRECT SENSORS—PRINCIPLES

For indirect sensors, a miniature transducer is attached to the distal tip of one or two fibers, as shown in Fig 7.5, and is, in effect, a miniature chemical/physics laboratory. It was suggested (Lubbers and Opitz, 1975) that such an optical transducer should be called an *optode*, similar to the term electrode used in electricity.[11] The input light sent through the fiber actuates the transducer, which in turn interacts with the sample. The output light sent back from the transducer is again analyzed by an optical processing system. If the optode is physical in nature, the parameters that can be measured are physical, such as temperature or pressure. If it is chemical, the parameters are chemical, such as pH or glucose content in the blood. Each category will be discussed separately using several examples.

7.5.1 Indirect Physical Sensors

Various indirect sensors have been developed for measuring either fluid pressure (P) or tissue temperature (T), blood velocity, or blood flow. These are based on different principles and on different optodes, some of which are described below. The fluorescence of the optode (Fig. 7.6a), the reflection from the optode (Fig. 7.6b), or the transmission through the optode (Fig. 7.6c) may be used for physical measurements on biological samples.

---

[11] Some people call it optrode, but in the opinion of the author the letter *r* should be omitted.

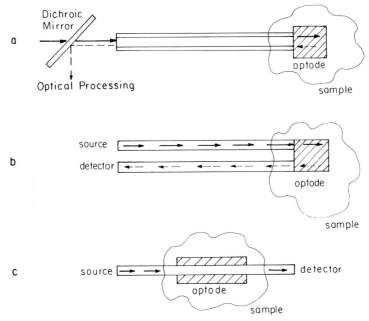

FIGURE 7.5 Indirect fiberoptic measurements.

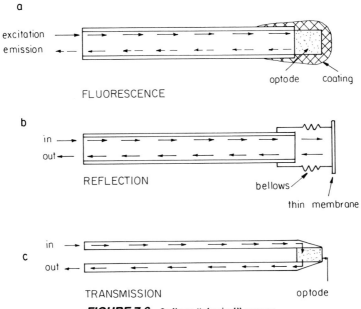

FIGURE 7.6 Indirect "physical" sensors.

7.5.1.1 Pressure Sensors

A common method for measuring intravascular blood pressure uses an ordinary catheter filled with fluid that is connected to an external transducer. These are simple devices that have limited accuracy. The evolution of miniature semiconductor devices and integrated circuits has resulted in an interest in electronic biosensors. It was suggested that miniature electronic devices, such as a silicon strain gauge, could be attached to the tip of a catheter and used for measuring pressure in blood vessels. The major problem was that they were not miniature. They were also affected by electromagnetic interference and did not stand up to the hostile environment of the body liquids. Fiberoptic pressure sensors do not suffer from these problems and are more suitable for pressure measurement inside the body. Most of these sensors are based on the simple principle shown schematically in Fig. 7.7. A mechanical optode is attached to the distal tip of a fiber (or a bundle of fibers). The optode is inserted in a liquid, whose pressure is P. Light (e.g., laser light) is sent through a fiber and reflected from a thin membrane. At a "normal" pressure P_0, the membrane is flat; under these conditions, some light I_0 is reflected from the membrane. When the optode is exposed to higher pressure, the membrane is displaced or curved and a different amount of light is reflected back. There is a correlation between the amount of light reflected through the fiber and the pressure P of the liquid. The reflected light can be measured by a photodetector and this value is then related to the pressure.

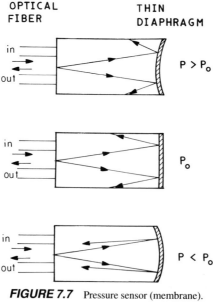

FIGURE 7.7 Pressure sensor (membrane).

The past two decades have witnessed several reports on the development of miniature optodes that serve as pressure transducers for biomedical applications. Most of these transducers were aimed at measuring blood pressure inside arteries, which meant that the outer diameter of the optode was limited to a few mm (<3 mm). The major difference between the various devices is the design of the pressure transducer (Lindstrom, 1970; Matsumoto and Saegusa, 1978; Lawson and Tekippe, 1983). Some transducers were based on the displacement of a mirror and some on the bending of a membrane. Hansen (1983) produced a more sophisticated pressure sensor in which a thin (10 μm) membrane was connected to the distal tip of a bifurcated fiberoptic bundle. The signal obtained in this manner varied linearly with the pressure in the range of 100 to 300 mm Hg and exponentially between 300 and 3000 mm Hg. The temperature drift of the instrument was acceptable and the time response was better than 1 msec.

Several experiments have used the fiberoptic pressure sensor *in vivo*; the results correlated well with conventional measurements. Clinical experiments have also been done to monitor blood pressure as well as pressure in the bladder, urethra, and rectum. Uterine pressure can be monitored during labor. These experiments demonstrate the feasibility of using fiberoptic sensors clinically (Wlodarczyk, 1989).

7.5.1.2 Temperature Sensors

Temperature sensors widely used in biomedical applications are usually electrical. Some are based on a pair of chemically different metal wires that are joined together in a point (thermocouple). An electric current is generated that depends on the temperature of the junction between metals. Other temperature sensors are based on a miniature thermistor whose electrical resistance varies with temperature. A thin thermocouple or a miniature thermistor has been routinely placed at the end of a catheter and external electronic circuitry has been used to measure the tip temperature. Such thermometers are well established, reliable, and fairly inexpensive. Because they absorb the laser radiation, they are not appropriate for real-time temperature measurements with laser-treated tissue. They are also not useful in applications where strong electromagnetic interference is present. Hyperthermia is discussed next in some detail in order to explain one of the important uses of fiberoptic temperature sensors.

Hyperthermia using Radio Frequency or Microwave

It is well established that controlled heating of tumor tissue to certain temperature levels (i.e., hyperthermia) leads to tumor regression. In a typical treatment, the temperature of the tumor is elevated to 42.5–43.5°C, for a period of 20 to 60 min. Heat-generating equipment with localized deposition of energy into the tissue is needed, as well as a thermometry system that allows accurate measurement of tissue temperature in space and time. Local hyperthermia may also work in synergism with chemotherapy, radiation therapy, and photodynamic therapy (PDT).

Local heating may be based on the absorption of ultrasound waves at a frequency of 1–10 MHz in tissue. Ultrasound waves are strongly focused to heat a small volume within the tumor and the focal spot is scanned in such a way that the temperature rise in the whole tumor is uniform. An Nd:YAG laser heating method has already been discussed in Section 3.7.1.1 and endoscopic laser hyperthermia will be discussed in Section 9.7.5. An alternative method is based on the absorption of electromagnetic (EM) waves in biological tissue. Radio-frequency (RF) waves at 13.56 or 27.13 MHz and microwaves (MW) at 2450 MHz are generally used because of strong absorption in tissue at these frequencies. Both heating methods are being used in preclinical and clinical studies. Efficient cancer treatment depends on fast and accurate measurement of the temperature in the tumor. The thermometer can also be used with a feedback system to control the output power of the heating system and provide the desired temperature field within the tumor. This task is particularly difficult when heating is carried out by RF or MW fields. The traditional thermocouple or thermistors contain metallic conductors which cause two problems: (i) regions of local heating in the probe and the surrounding tumor due to the presence of the conductor and (ii) electromagnetic field induction of currents and voltages in the metallic conductors, resulting in erroneous temperature readings.

These problems can be alleviated by using fiberoptic indirect sensors that are nonmetallic and fairly immune to electromagnetic noise. These are based on optodes in which some physical property changes as a function of temperature. A large number of optodes have been tried in the laboratory, but only a few have been tried *in vivo* and an even smaller number are available commercially. A few examples will be discussed to illustrate the operating principles of such sensors. These are not necessarily the sensors that will eventually evolve as the "best."

(i) Liquid crystal optodes: The reader is probably familiar with liquid crystals (LCs) that are widely used in electronic displays, such as in digital watches or pocket calculators. These are materials that show a dramatic change in color in response to a temperature change. LCs are used in inexpensive (or disposable) thermometers, in which the temperature is shown by colored digits. These LCs were attached to the distal end of an optical fiber and used in the same manner for measuring temperatures. In the experiments of Rozzell *et al.* (1974), red light (670 nm) from an LED was transmitted through a fiber bundle onto a layer of cholesterol compound LC. The reflected light intensity was dependent on temperature. This fiberoptic sensor was tried *in vitro* in the presence of a microwave field. The response time was about 4 sec and the temperature resolution was about 0.1°C. The complexity of the LC probes tends to limit their widespread application.

(ii) Luminescent probe: The luminescence of many materials depends strongly on temperature T, in some range. In principle, by measuring the luminescence emission intensity, T can be determined. The luminescent material (e.g., lanthanum oxysulfide activated with europium) is attached to the distal tip of a optical fiber. UV radiation of intensity I_i is sent through the fiber to excite lumi-

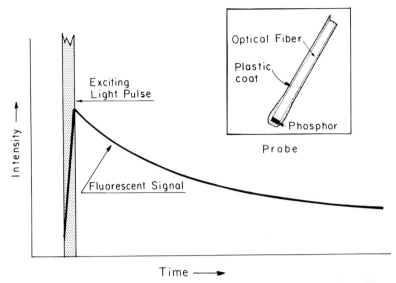

FIGURE 7.8 Commercial temperature sensor—decay time depends on T. (Courtesy of Luxtron.) (Insert) T sensor probe.

nescence. The emitted light of intensity I_o is transmitted back through the same fiber and its intensity related to the temperature. This method has the disadvantage that the incident intensity I_i must be monitored continuously to make sure that it is constant. To overcome this problem, another type of luminescent material is used. Such a material is excited by pulses of UV light and emits visible light that decays with a characteristic time constant τ, as shown in Fig. 7.8. There are materials (e.g., magnesium fluorogermanate activated with manganese) in which τ is linearly dependent on T and thus τ can be measured and T determined. This method is independent of I_i and easier to calibrate. Fiberoptic sensors that are based on luminescence have been successfully tested *in vivo* and are available commercially (Sun *et al.*, 1989). The luminescent sensor was designed to work in the presence of strong EM fields. Monitoring temperatures in hyperthermia cancer treatment has already been clinically tried. In such sensors, the temperature resolution is 0.1°C and the time response is about 1 sec.

7.5.1.3 Flow Measurement

Section 7.4.2.2 described the thermodilution technique for measuring blood flow. This method is based on injection of a cold fluid into the blood and monitoring the temperature of the blood downstream. It is assumed that in a very short period of time there will be little exchange of heat between the cold fluid and the blood. As the fluid flows with the blood, the mixture of blood and fluid is colder than the blood and warmer than the fluid. One can calculate the relation of the temperature of the mixture T_m to the temperatures of the cold fluid T_f and the temperature of the blood T_b. If all the parameters are known, measurements of T_m are sufficient to determine the blood flow.

This method has several advantages:

(i) Blood sampling is not required.
(ii) The cold fluid (saline solution or 5% dextrose in water) is relatively harmless.
(iii) Calibration is simple.
(iv) Measurements can be repeated at short intervals.
(v) Either regional or total blood flow can be measured.

In the conventional thermodilution method, thermistors or thermocouples are used. Because they are immune to EM interference and have faster responses time, fiberoptic sensors are preferable for this application.

The thermodilution method can be performed conveniently using a catheter that includes a fiberoptic temperature sensor. The distal tip of the sensor (which includes the optode) is placed at the tip of the catheter. The catheter is inserted at the relevant point, such as the coronary sinus. Cold fluid is injected continuously through the catheter and the fiberoptic sensor measures the temperature T_m of the blood stream a few centimeters from the point where the fluid was injected. This method has been demonstrated *in vivo*, (De Rossi *et al.*, 1980), in open chest experiments. The results compared favorably to those obtained with a standard EM flowmeter which was employed in the same experiment. This experiment showed the feasibility of using fiberoptic sensors, in conjunction with catheters or endoscopes, for flow measurements. These sensors may be quite useful in terms of integrated fiberoptic systems.

7.5.2 Indirect Chemical Sensors

In this category, there are several sensors for measuring pH, blood gases, glucose content, and the content of other chemical substances in the blood. Typical "chemical" indirect sensors are shown in Fig. 7.9. Some of the principles of operation of the various chemical sensors are discussed below (Seitz, 1984; Gottlieb *et al.*, 1991).

7.5.2.1 pH Sensors

Conventional methods of measuring pH in the body are based on the measured potential of microelectrodes. Fiberoptic pH probes offer many advantages over these electrodes, as mentioned earlier; they are potentially safer (no electrical connections), more flexible, cheaper, and disposable. An optode is attached to the distal tip of the fiber and is coated with a thin protective coating which permits small ions, such as hydrogen ions, to penetrate and interact with the optode but keeps larger ions out. When the whole tip is inserted into blood, the interaction with the optode, through the coating, gives rise to optical changes which are mea-

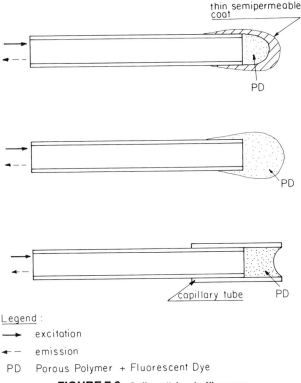

Legend :
→ excitation
←– emission
PD Porous Polymer + Fluorescent Dye

FIGURE 7.9   Indirect "chemical" sensors.

sured through the fiber. This method is illustrated with two examples: one is based on changes in the optical absorption of an indicator and the other on changes in the luminescence properties.

Optical Absorption Method for Measuring pH

A common technique for determining pH is based on substances called indicators, which are used to indicate the acidity or alkalinity of a solution by changing color. Such a color change can be observed through an optical fiber. One of the first pH sensors was based on a dye, phenol red, as an indicator (Peterson *et al.*, 1980). It was found that the dye changes color depending on the pH, in the physiological range of interest (pH between 7.0 and 7.4). This effect can be quantified by measuring the optical absorption of the dye. Two thin fibers are inserted into a hollow catheter and placed a few millimeters away from the catheter tip. The dye is attached to polymer gel microspheres of 10 μm diameter. These spheres are mixed with smaller spheres, about 1 μm in diameter; packed near the tips of the fibers, at the end of the catheter; and encapsulated by a thin membrane. Light sent through the input fiber is scattered by the tiny spheres. Part is absorbed by the dye

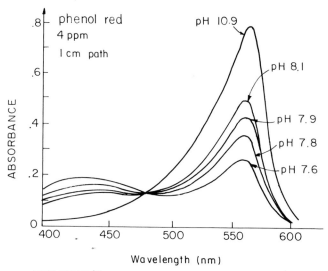

FIGURE 7.10   Determination of pH of blood by absorption.

and the rest is transmitted back through the output fiber. The smaller spheres assist the scattering process. The amount of light transmitted back depends on the absorption in the dye. Two irradiating colors, red (600 nm) and green (550 nm), from two LEDs are used alternately. The green light is absorbed by the dye and the amount transmitted is measured by a small semiconductor photodetector. This amount depends on the pH, as shown in Fig. 7.10. The red light is not pH dependent, serving only to calibrate the system. The ratio I(green)/I(red) is a direct measure of the pH. The whole fiberoptic sensor is simple in concept and can be quite small and battery operated.

The fiberoptic pH probe has been assessed *in vitro*, in solutions of known pH, and has been found to be accurate within 0.01 pH unit. The signal is temperature dependent, changing by about 0.02 pH unit per degree. The response time of the system is approximately 45 sec. The same probe has also been assessed *in vivo* in animal experiments. The pH of the blood was varied by using different CO_2 levels in the inspired air. The fiberoptic probe was inserted into the jugular vein of the animal and used to measure pH. The results compared favorably with those obtained by drawing blood samples and using conventional techniques to determine pH. A modified version of the same probe can also be used to monitor the pH of tissue. A probe incorporated inside a hypodermic needle has been used to measure the pH in the heart wall of a dog. The use of this probe is not limited to the cardiovascular system or to the pH range 7.0–7.4. Similar probes could be used in the gastrointestinal system, where the pH range of interest is 2–8. This type of fiber sensor may be inserted into the gastrointestinal tract via catheters or endoscopes. Because of its thinness, it may even be used to monitor the pH for an extended period of time, with minimum discomfort of the patient.

Luminescent pH Sensor

This method is based on a luminescent dye whose luminescence is dependent on pH. One example (Gehrich *et al.*, 1986) is a water-soluble dye (hydroxypyrene trisulfonic acid) that has two forms: acidic and basic. Both forms can be excited to emit a luminescence that peaks at 520 nm. There is a marked difference between the excitation spectra of the two forms, as shown in Fig. 7.11. The excitation spectrum peaks at 410 nm for the acidic form and at 460 nm for the basic form. Therefore, the luminescence intensity at 520 nm is excited once at 410 nm and again at 460 nm. The ratio of the results gives the relative concentrations of basic and acidic forms of the dye, which is then related to the pH of the solution.

For fiberoptic sensors, the dye can be bound to a hydrophilic polymer material and coated with a thin overcoat layer. Light at the two wavelengths (410 and 460 nm) is sent alternately through a single fiber to the optode. The luminescence at 520 nm is detected via the same fiber and is measured by a photodetector. The pH value is calculated electronically from these results.

The fiberoptic probe has been tested *in vitro* and the results obtained in the pH range 6.7–7.9 agreed with the results obtained with a conventional pH meter.

7.5.2.2 Pco₂

The dyes used to measure pH can also be used to measure P_{CO_2}, the partial pressure of CO_2 in the blood. The pH of a bicarbonate solution depends on the P_{CO_2} with which it is in equilibrium. By measuring the pH of the solution, the P_{CO_2} can be determined (Vurek, 1984; Geherich *et al.*, 1986). Usually, an increase in P_{CO_2} will reduce the pH of the bicarbonate solution, causing a decrease in the

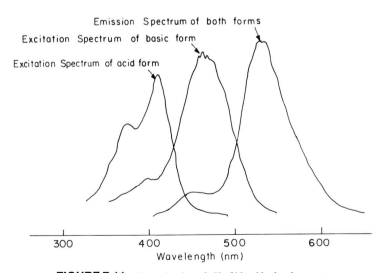

FIGURE 7.11 Determination of pH of blood by luminescence.

luminescence of the dye. The major difference between this optode and the pH optode is that ions are excluded from the probe; only gas molecules are allowed to enter. In some optodes, this is achieved by encapsulating the bicarbonate solution in a silicone matrix that provides the necessary isolation from the ions in the blood. There is still an overcoat of thin opaque cellulose layer on the optode to protect it and its optical system from external light. By choosing the correct concentration of bicarbonate and dye, the P_{CO_2} can be determined in the range $10-1000$ mm Hg with 1% accuracy. Sensors which are based on polymer-immobilized, pH-sensitive, fluorescent dyes were discussed by Munkholm *et al.* (1988).

7.5.2.3 Po$_2$

In the previous measurements, luminescence emission was directly proportional to a specific parameter. There are cases in which the luminescence is inversely proportional to a specific parameter. A well-known case is the quenching of luminescence by oxygen. In certain materials that luminesce, the luminescence intensity decreases as the partial pressure of oxygen increases. This effect can be utilized to sense the partial pressure of oxygen in the blood.

Several dyes have been mentioned in the literature, (Lubbers and Opitz, 1983; Vurek, 1984; Wolfbeis and Leiner, 1988) for determining Po$_2$ in the range $20-150$ mm Hg. The dye is embedded in a polymer and attached to the distal tip of a fiber. The polymer is normally coated with a thin layer of gas-permeable material. In the early models, the exciting light was in the UV, thus requiring fused silica fibers for transmission. Using dyes that can be excited by blue light, newer sensors are based on glass (or even plastic) fibers. With commercial instruments, two dyes are used. The luminescence of one of the dyes is insensitive to oxygen pressure and serves to calibrate the whole system. The other dye serves for oxygen monitoring.

7.5.2.4 Glucose

A fiberoptic sensor that may be easily inserted into the blood vessels and even kept in the body for some time may be useful as part of a control system of an insulin dispenser for diabetics. The fiberoptic glucose sensor is based on a method (Schultz *et al.*, 1982), known as the competitive binding method, that may have wide application. It is shown schematically in Fig. 7.12.

The optode consists of a hollow tube plugged at one end, with its walls freely permeable to glucose. The tube is filled with a fluorescent dye, fluorescein, which is bound to a soluble glucose polymer. The inside of the hollow fiber is coated with concanavalin A, a material which binds glucose. A bifurcated fiber bundle is inserted into the open end of the hollow fiber and transmits light into the optode and back again.

In the absence of glucose, the binding compound binds the fluorescence polymer to the wall. The luminescence takes place in a geometric location that is removed from the field of view of the fiber bundle. When the optode is inserted into

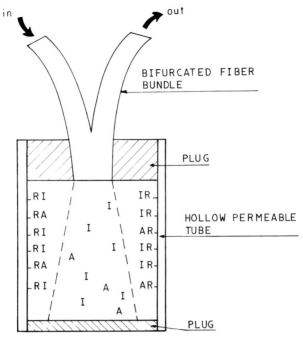

FIGURE 7.12   Glucose sensor. R, reagent (e.g., con A); A, analyte (e.g., glucose); I, indicator (e.g., fluorescine-labeled dextron).

a glucose solution, the glucose penetrates the optode and displaces the polymer from the walls. The fluorescein concentration in the solution (away from the walls) increases. Under excitation, there is more luminescent light which the fiber bundle can transmit to the outside. Therefore, the more glucose in the solution, the higher the luminescence observed in the sensor. This type of fiberoptic sensor may well be a prototype for a generic class of sensors to detect and analyze other blood components.

7.6 BIOCHEMICAL FIBEROPTIC SENSORS—PRINCIPLES

Special sensors are needed to detect chemicals, toxins, and other substances in complex solutions such as blood. In biochemistry, the interactions of antibody and antigen, for example, have proved to be highly specific and extremely useful. Biochemical fiberoptic sensors are based on optodes that rely on such interactions and are therefore highly specific. One type of biochemical sensor that makes use of evanescent wave spectroscopy (Andrade et al., 1985; Sutherland et al., 1984) is discussed below.

In Section 4.4 we discussed the total internal reflection of light (see Fig. 4.6). Light strikes the interface between two transparent media. Impinging on the me-

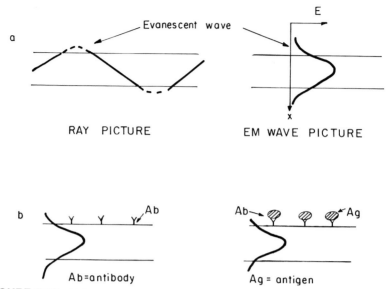

FIGURE 7.13 (a) Evanescent waves—geometrical (ray optics) and physical optics pictures. (b) Evanescent wave biochemical sensor.

dium having the lower index of refraction, it is reflected back into the medium of high index of refraction. The light, however, penetrates a distance d, of the order of a wavelength of light, into the medium of lower index of refraction. The ray picture is shown schematically in Fig. 7.13a. This phenomenon may also be described as propagating fields inside the medium of high refractive index accompanied by a flow of energy. In the medium of low index, there is no flow of energy but rather a nonpropagating field called an evanescent wave, which decays exponentially with distance from the surface. This is also shown in Fig. 7.13a.

Let us now consider an unclad optical fiber which consists of glass core only, without any glass cladding layer. The fiber may be coated by a thin layer of polymer. One may also choose a polymer which fluoresces in the visible when exposed to UV. If UV light is sent through the fiber, the evanescent wave excites fluorescence in the polymer coating. It has been shown that, under certain optical conditions, the visible fluorescence is coupled back into the core and is transmitted back through the same optical fiber, where it may be easily detected.

A similar situation occurs when the unclad fiber is inserted into a solution, i.e., dye, that fluoresces in the UV. The UV light is again coupled into the input end of a fiber. The visible fluorescence, excited by the evanescent wave in the solution, is transmitted back through the fiber to the input end. The presence of luminescent species in a solution may thus be detected optically. The fluorescence intensity may serve to determine the concentration of the solute in the solution.

Suppose now that the concentration of some antigen (Ag) is to be determined in a complex solution that contains many chemicals. Also suppose that an anti-

TABLE 7.1 Fiberoptic Sensors—Advantages and Problems

Advantages	Problems
Miniature, rugged and biocompatible probes	Optode construction
Specific, sensitive, and accurate	Calibration
Several analytes can be measured	Small signal
Easily inserted into the body	Deterioration of optodes (aging)
Continuous monitoring	Response time (too slow)
Inexpensive and portable system	Limited dynamic range
Disposable probes	

body (Ab) against the antigen is available and that this antibody can be covalently bonded on the surface of an unclad fiber. Let us also assume that both antigen Ag and antibody Ab are fluorescent. At first the antibody is attached to the tip of the unclad fiber (it is called an immobilized Ab). When UV light is sent through the fiber, the evanescent wave excites fluorescence that is transmitted back through the fiber. When the fiber tip is inserted into the complex solution (i.e., blood), the evanescent wave will penetrate only a short distance into the solution; it will generally not excite an additional fluorescence. If, however, the specific antigen is present, it will bind to the immobilized Ab. The evanescent wave will thus excite the Ag and additional fluorescence will be observed. The intensity of the additional fluorescence is proportional to the concentration of Ag in the solution.

The method just described is one example of the potential of combining the selectivity offered by a biochemical technique and the sensitivity offered by optical techniques. Fiberoptic chemical sensors based on biochemical optodes will be used to detect toxins, drugs, and other biochemicals in the blood (Kulp *et al.*, 1987; Tromberg *et al.*, 1987; Arnold, 1985).

Some of the advantages and the problems of optical fiber sensors, in general, are summarized in Table 7.1

7.7 DIRECT AND INDIRECT FIBEROPTIC SENSORS—ADVANCES

7.7.1 Laser Doppler Velocimetry (LDV)

This type of sensor is illustrated with examples related to measurements of blood flow. For example, cardiologists need to measure blood supply to the heart in order to know if the coronary arteries are blocked and to what degree. The problem is modeled on the transmission of fluid from a container through a straight pipe. The fluid consists of a large number of particles, each of which moves with velocity v, measured in cm/sec. Within a small volume, each particle moves in a different direction with a different velocity. If the fluid is stationary, the average velocity of all the particles is $V = 0$. If the liquid is flowing, the average velocity of all the particles is V. The fluid flows through the tube and emerges at the end. What is the flow? It is the *volume* of fluid that passes through the end of the tube per unit time. The flow F is measured in cubic centimeters per second. The flow is determined by averaging the velocities of all the particles in the fluid.

Blood is, however, thicker than water, and its flow characteristics are more complex. It flows into blood vessels which are not rigid straight tubes, and the flow is pulsatile. Blood is a complex liquid, and when it flows through the vessels (especially if they are blocked or constricted) it can become turbulent. In short, mathematical models which can be applied to water are less reliable when applied to blood and determining blood flow is not easy. Fiberoptic sensors offer several methods for measuring the blood velocity and/or flow.

When a wave of frequency f and velocity c impinges on a stationary object, it is reflected at the same frequency. If the object moves with a velocity v, the reflected frequency f' is different from f. This is the Doppler effect, and the frequency difference $\delta f = f - f'$ (called Doppler shift) is given by the formula

$$\delta f / f = 2v/c \qquad (7.1)$$

An example of the Doppler effect is the whistle of a train. If the train is moving toward an observer, the whistle sounds as if it has a higher pitch than if the train is moving away. As the train passes the observer, the pitch of the whistle falls. This does not involve a reflected wave, but rather a sound wave generated on the moving object itself, but the effect is the same. If the train is stationary (its velocity $v = 0$), the pitch emitted by the whistle is constant. If the train moves, the frequency f' (pitch) that is heard depends on the velocity v of the train. If both f and f' are known, Eq. (7.1) can be used to calculate the velocity v. This is true for trains, but it is also true for electromagnetic waves. A radar beam of frequency f (of order 10^{10} Hz), transmitted toward a moving object (e.g., an airplane or a tennis ball), produces a reflected beam with a different frequency f'. The Doppler shift will again be given by the formula (7.1); if δf is measured, the velocity v of the object can be determined. The same principle applies to beams of laser light which are reflected from moving objects. This method is called laser Doppler velocimetry (LDV) and it provides accurate measurements of moving objects. LDV has been introduced as a method for measuring the speed of cars by police forces in several countries.

Clearly, LDV can be used with fiberoptic systems. The only difference is that the incident and reflected laser beams are sent through optical fibers, rather than through air. The method is then called fiberoptic LDV or FOLDV. The laser beam interacts with a small volume of blood near the distal end of the two fibers. If there is a distribution of velocities of blood cells in this volume, the LDV will reveal information about both the velocity distribution and the number of particles moving at each velocity. The sum over all the velocities can be integrated and the flow obtained.

Laser Doppler velocimetry in its fiberoptic version is potentially useful in medicine, especially in cardiology. If a cardiologist is interested in determining the blood flow in a diseased coronary artery, an ultrathin angioscope can be threaded into the artery and positioned next to the blockage under observation. An FOLDV measurement is carried out as described above. The signal obtained from the detector reveals the velocity distribution near the tip of the angioscope. The physician can easily move the distal end of the angioscope from the center of the coronary artery to the sides to measure velocity there. As mentioned above, blood flow can be calculated from these results. It is possible to determine whether the blood flow is laminar (smooth flow in one direction) or turbulent. Using this information, the cardiologist can decide if the blockage must be removed. If so, FOLDV measurements can be repeated to determine whether sufficient blood flow has resumed.

The LDV technique is simple in concept and easy to understand. However the construction of a reliable fiberoptic system for this purpose is far more complex. LDV may easily be used to measure changes in flow, but it is difficult to calibrate the absolute flow measurements and the angioscope itself is large enough to alter the flow substantially during LDV.

The first experiments using LDV for blood flow were reported almost two decades ago (Tanaka and Benedek, 1975). Over the years, there has been slow progress in this field. More recently, there have been several reports on the successful use of fiberoptic LDV *in vivo*, both externally and endoscopically. Nilsson *et al.* (1980) used fiberoptic LDV to measure the blood flow in tissue. For this purpose, they used an HeNe laser beam which penetrates about 1 mm into the skin. The HeNe laser beam is scattered by the red blood cells which flow in the capillaries in the skin. This scattered light can be collected by another fiber and transmitted back to the LDV system. The signal obtained in this measurement is dependent on the geometry of the vascular bed in the skin, the concentration of the red blood cells, and their velocity distribution. Nilsson *et al.* showed the feasibility of using the fiberoptic LDV technique for measuring changes in blood flow in the skin as a function of skin cooling or breath holding. In principle, this is a useful method which may result in a continuous measurement of tissue blood flow anywhere on the surface of the body. In practice, it may be useful only for measuring changes in blood flow. Smits, Aarnoudse, and their colleagues (1986) used a similar fiberoptic LDV system for measuring the blood flow in the scalp of a fetus during labor.

Fiberoptic LDV measurements have been further improved during the last few years. Kilpatrick *et al.* (1988) used FOLDV techniques to measure blood velocities *in vivo* in the coronary sinus in anesthetized dogs. They compared the FOLDV results to those obtained with electromagnetic flow probes and found good correlation. More important, by moving the distal tip of the fiber, a velocity profile across an arterial stenosis (narrowing) was measured.

References

Alfano, R. R., and Alfano, M. A. (1985). Medical diagnostics: a new optical frontier. *Photonics Spectra* **December**, 55–60.

Andersson-Engels, S., Johansson, J., Svanberg, S., and Svanberg, K. (1989). Fluorescence diagnosis and photochemical treatment of diseased tissue using lasers: part I. *Anal. Chem.* **61**, 1367a–1373a.

Andersson-Engels, S., Johansson, J., Svanberg, S., and Svanberg, K. (1990). Fluorescence diagnosis and photochemical treatment of deseased tissue using lasers: part II. *Anal. Chem.* **62**, 19a–27a.

Andrade, J. D., Vanwagenen, R. A., Gregonis, D. E., Newby, K., and Lin, J. N. (1985). Remote fiber optic biosensors based on evanescent-excited fluoro-immunoessay: concept and progress. *IEEE Trans. Electron. Devices* **ED-32**, 1175–1179.

Arnold, M. A. (1985). Enzyme based fiber optic sensor. *Anal. Chem.* **57**, 565–566.

Callahan, S. (1977). Moth and candle: the candle flame as a sexual mimic of the infrared wavelengths from a moth sex scent (pheromone). *Appl. Opt.* **16**, 3089–3097.

Culshaw, B. (1982). *Optical Fibre Sensing and Signal Processing*, London: Peter Peregrinus.

De Rossi, D., Benassi, A., L'Abbate, A., and Dario, P. (1980). A new fibre-optic liquid crystal catheter for oxygen saturation and blood flow measurements in the coronary sinus. *J. Biomed. Eng.* **2**, 257–264.

Gehrich, J. L., Lubbers, D. W., Opitz, N., Hansmann, D. R., Miller, W. W., Tusa, J. K., and Yafuso, M. (1986). Optical fluorescence and its application to an intravasclar blood gas monitoring system. *IEEE Trans. Biomed. Eng.* **BME-33**, 117–131.

Gottlieb, A., Divers, S. and Hui, H. K. (1991). In vivo applications of fiberoptic chemical sensors. In: Wise, D. L., and Wingard, L. B. (Eds.), *Biosensors with Fiberoptics*. Clifton, NJ: Humana Press.

Hall, E. A. H. (1988). Recent progress in biosensor development. *Int. J. Biochem.* **20**, 357–362.

Hansen, T. E. (1983). A fiberoptic micro-tip pressure transducer for medical applications. *Sensors Actuators* **4**, 545–554.

Kapany, N. S., and Silbertrust, N. (1964). Fibre optics spectrophotometer for in vivo oximetry. *Nature* **204**, 138–142.

Katzir, A., Bowman, H. F., Asfour, Y., Zur, A., and Valeri, C. R. (1989). Infrared fibers for radiometer thermometry in hypothermia and hyperthermia treatment. *IEEE Trans. Biomed. Eng.* **36**, 634–636.

Kilpatrick, D., Kajiya, F., and Ogaswara, Y. (1988). Fibre optic Doppler measurement of intravascular velocity. *Austral. Phys. Eng. Sci. Med.* **11**, 5–14.

Korb, H., Bock, J., Hoeft, A., and DeVivie, R. (1989). Determination of central blood volume and extravascular lung water by a double fiberoptic device. *Proc. SPIE* **1967**, 69–74.

Kulp, T. J., Camins, I., Angel, S. M., Munkholm, C., and Walt, D. R. (1987). Polymer immobiized enzyme optrodes for the detection of penicillin. *Anal. Chem.* **59**, 2849–2853.

Lawson, C. M., and Tekippe, V. J. (1983). Fiber-optic diaphragm-curvature pressure transducer. *Opt. Lett.* **8**, 286–288.

Lindstrom, L. H. (1970). Miniaturized pressure transducers intended for intravascular use. *IEEE Trans. Biomed. Eng.* **BME-17**, 207–215.

Lubbers, D. W., and Opitz, N. (1975). Die pCO2 pO2 optode: eine neue pCO2 bzw. pO2 messonde zur messung des pCO2 oder pO2 vas gasen und flussigkeiten. *Z. Naturforsch.* **30C**, 532–533.

Lubbers, D. W., and Opitz, N. (1983). Optical fluorescence sensors for continuous measurement of chemical concentrations in biological systems. *Sensors Actuators* **4**, 641–654.

Mandoli, D. F., and Briggs, W. R. (1985). Fiber optics in plants. *Sci. Am.* **202**, 80–88.

Martin, M. J., Wickramasinghe, Y. A. B. D., Newson, T. P., and Crowe, J. A. (1987). Fibre optics and optical sensors in medicine. *Med. Biol. Eng. Comput.* **25**, 597–604.

Matsumoto, H., and Saegusa, M. (1978). The development of a fibre optic catheter tip pressure transducer. *J. Med. Eng. Technol.* **2**, 239–242.

Mayevsky, A., and Chance, B. (1982). Intracellular oxidation reduction state measured in situ by a multichannel fiber optic surface fluorometer. *Science* **217**, 537–540.

Milanovich, F. P., Hirschfeld, T. B., Wang, F. T., Klainer, S. M., and Walt, D. R. (1984). Clinical measurements using fiber optics and optrodes. *Proc. SPIE* **494**, 18–24.

Munkholm, C., Walt, D. R., and Milanovich, F. P. (1988). A fiber-optic sensor for CO_2 measurement. *Talanta* **35**, 109–112.

Nilsson, G. E., Tenland, T., and Oberg, P. A. (1980). Evaluation of a laser Doppler flowmeter for measurement of tissue blood flow. *IEEE Trans. Biomed. Eng.* **BME-27**, 597–604.

Peterson, J. I., Goldstein, S. R., and Fitzgerald, R. V. (1980). Fiber optic pH probe for physiological use. *Anal. Chem.* **52**, 864–869.

Peterson, J. I., and Vurek, G. G. (1984). Fiber-optic sensors for biomedical applications. *Science* **224**, 123–127.

Pini, R., Salmbeni, R., Vannini, M., Cavalieri, S., Barone, R., and Caluser, C. (1989). Laser dentistry: Root canal diagnosis techniques based on UV-induced fluorescence spectroscopy. *Laser Surg. Med.* **9**, 358–361.

Polanyi, M. L., and Hehir, R. M. (1962). In vivo oximeter with fast dynamic response. *Rev. Sci. Instrum.* **33**, 1050–1054.

Renault, G. (1987). Clinical applications of laser fluorometer. *Lasers Optronics* **December**, 56–59.

Renault, G., Sinet, M., Muffat-Joly, M., Cornillault, J., and Pocidalo, J. (1985). In situ monitoring of myocardial metabolism by laser fluorimetry: relevance of a test of local ischemia. *Laser Surg. Med.* **5**, 111–122.

Rozzell, T. C., Johnson, C. C., Durney, C. H., Lords, J. L., and Olsen, R. G. (1974). A nonperturbing temperature sensor for measurements in electromagnetic fields. *J. Microwave Power* **9**, 241–249.

Schultz, J. S., Mansouri, S., and Goldstein, I. J. (1982). Affinity sensor: a new technique for developing implantable sensors for glucose and other metabolites. *Diabetes Care* **5**, 245–253.

Seitz, W. R. (1984). Chemical sensors based on fiber optics. *Anal. Chem.* **56**, 16–34.

Smits, T. M., Aarnoudse, J. G., and Zijlstra, W. G. (1986). Red blood flow in the fetal scalp during hypoxemia in the chronic sheep experiment: a laser Doppler flow study. *Pediatr. Res.* **20**, 407–410.

Sun, M. H., Wickersheim, K. A., and Kim, J. (1989). Fiberoptic temperature sensors in the medical setting. *Proc. SPIE* **1967**, 15–21.

Sutherland, M., Dahne, C., Place, J. F., and Ringrose, A. S. (1984). Optical detection of antibody antigen reactions at a glass liquid interface. *Clin. Chem.* **30**, 1535–1538.

Tanaka, T., and Benedek, G. B. (1975). Measurement of the velocity of blood flow (in vivo) using a fiber optic catheter and optical mixing spectroscopy. *Appl. Opt.* **14**, 189–196.

Teng, P., Nishioka, N. S., Anderson, R. R., and Deutsch, T. F. (1987). Optical studies of pulsed-laser fragmentation of biliary calculi. *Appl. Phys.* **B 42**, 73–78.

Tromberg, B. J., Sepaniak, M. J., VoDinh, T., and Griffin, G. D. (1987). Fiber optic chemical sensors for competitive binding fluoroimmunoassay. *Anal. Chem.* **59**, 1226–1230.

Volz, R. J., and Christensen, D. A. (1979). A neonatal fiberoptic probe for oximetry and dye curves. *IEEE Trans. Biomed. Eng.* **BME-26**, 416–422.

Vurek, G. G. (1984). In vivo optical chemical sensors. *Proc. SPIE* **494**, 2–6.

Walt, D. R. (1992). Fiberoptic sensors for continous clinical monitoring. *Proc. IEEE* **80**, 903–911

Wise, D. L., and Wingard, L. B. (1991). *Biosensors with Fiberoptics*. Clifton, NJ: Humana Press.

Wlodarczyk, M. T. (1989). Dual wavelength catheter type fiber optic pressure sensor. *Proc. SPIE* **1067**, 8–13.

Wolfbeis, O. S. (1987). Fibre optic sensors in biomedical sciences. *Pure Appl. Chem.* **59**, 663–672.

Wolfbeis, O. S. (1992). *Fiber Optic Chemical Sensors and Biosensors*, Vols. I and II. Boca Raton, FL: CRC Press.

Wolfbeis, O. S., and Leiner, M. J. P. (1988). Recent progress in optical oxygen sensing. *Proc. SPIE* **906**, 42–48.

8

Fiberoptic Laser Systems for Diagnostics and Therapy

8.1 INTRODUCTION

An exciting aspect of lasers in medicine involves the integration of lasers and fiberoptic systems into medical tools. These systems make it possible to perform imaging, diagnostics, and therapy inside the body less invasively than with conventional methods. In the past, optical fibers were limited in their power transmission capabilities and were used mainly for illumination or for image transmission. These fibers were often used for constructing endoscopes that incorporated mechanical devices such as forceps; the optical fibers themselves were not commonly used for diagnostic or therapeutic applications. The situation changed with the development of high-quality fibers. It is now possible to send laser beams of relatively high power through glass fibers that can be incorporated into catheters or endoscopes; this laser light can be used to perform endoscopic laser therapy inside the body. This is augmented by the fact that power-transmitting fibers are now available not only for visible light but also for UV and IR radiation. These spectral regions are useful in laser therapy. This chapter discusses how lasers and fibers are coupled. The basics of laser–fiber integrated systems and their potential applications are discussed first, followed by a detailed discussion of important laser–fiber systems. Details of the clinical applications of all these systems are given in the following chapter.

An *integrated system* is composed of several systems that are coupled to operate as one unit. Each system is a functional unit that may be divided (again, conceptually) into more elementary parts called *subsystems*. Each of the subsystems consists of elementary building blocks called *components*. For example, a personal computer system is an integrated system. It may consist of systems such as the computer itself, a TV monitor, and a laser printer. The computer contains several subsystems such as the hard disc drive and floppy disc drives. The subsystems consist of mechanical components (e.g., knobs) or electronic components (e.g., transistors). The following parts are considered for laser–fiber integrated systems: (see also Table 8.4):

- *Integrated systems*: laser–fiber; laser–endoscope; laser–catheter.
- *Systems*: laser; endoscope; catheter.
- *Subsystems*: fiberscope; fiberoptic delivery unit.
- *Components*: fiber; nonordered fiber bundle; ordered fiber bundle; optical elements (lenses).

Earlier chapters described the components, subsystems, and systems, and some of these are briefly discussed below. This chapter concentrates on the integrated systems themselves and their functions. Potential applications are mentioned briefly in order to illustrate the uses of the integrated systems.

8.2 LASER–FIBER INTEGRATED SYSTEMS—FUNDAMENTALS

This book deals with the laser–fiber treatment of diseases that create localized diseased areas such as tumors, blood clots, or atherosclerotic plaque. Previous chapters discussed the scientific and medical aspects of disease that can be diagnosed and treated by lasers, fiberoptic imaging, and the laser therapeutic methods. This chapter tries to give a picture of the laser–fiber system as a complete unit that uses optical techniques for diagnosis and identification of diseased tissues. The system incorporates the various lasers necessary for treating each diseased area (Apfelberg, 1987; Artjushenko *et al.*, 1989).

The most general integrated system, the laser endoscope, consists of three systems: imaging, diagnosis, and therapy. Imaging is done by the fiberoptic endoscope system, which was described in Chapter 6. It may be enhanced either by use of special illumination techniques or by computer methods. Diagnosis is made using the fiberoptic sensor systems discussed in Chapter 7. The optical fiber used for diagnosis is inserted through one of the ancillary channels of the endoscope, and it facilitates diagnosis in the general location that is imaged by the endoscope. For therapeutic purposes, the laser–fiber system is used. Another fiber (power fiber) is inserted through an ancillary channel of the endoscope to carry the therapeutic laser beam. Both the imaging and the diagnostic techniques enable the physician to test the efficiency of the therapeutic procedure *in situ*. If imaging is not necessary, a catheter is used instead of an endoscope. The laser–fiber system and the laser catheter still provide diagnosis and treatment.

The operation of an integrated laser–fiber system may be illustrated by a series of hypothetical steps:

(i) If possible, find the exact location of the diseased area by using noninvasive imaging methods, such as x-ray fluoroscopy, CT, MRI, or ultrasound.

(ii) If (i) is not possible, insert a catheter or an endoscope through a natural opening, or through a hypodermic needle, while moving its distal end to the vicinity of the diseased area.

(iii) Determine the nature of the disease by using fiberoptic or nonoptical imaging and diagnostics and choose a course of therapy (heating, surgery, etc.).

(iv) Choose the most suitable laser and the mode of operation (short pulses, long pulses, CW, etc.), for treatment.

(v) Choose the optical fiber delivery system (i.e., fiber type, special tip).

(vi) Insert the optical fiber through the endoscope or the catheter and place it in the correct location.

(vii) Send a laser beam for therapy.

(viii) To ensure that there are no deleterious effects, control both the position of the fiber and the action of the laser during the procedure. This could again be performed by fiberoptic imaging or sensing.

(ix) At the end of the procedure use diagnostic or imaging techniques to assess the effect of the treatment or whether any damage has occurred.

8.3 COMPONENTS, SUBSYSTEMS, AND SYSTEMS—PRINCIPLES

Fiberoptic diagnostics, imaging techniques and laser therapy have been discussed earlier. In this section, emphasis is placed on the systems aspect. A description is given of the integrated systems that incorporate all the optical components, subsystems, and systems presented in earlier chapters. The important lasers which are used for diagnosis and therapy, the various optical fiber systems, and the special problems encountered when trying to deliver high power through fibers will be briefly discussed here and in the following sections.

8.3.1 Components—Optical Fibers

The optical properties of optical fibers, in general, and the transmission losses, in particular, were discussed in Section 4.8. The optical properties of materials at high power densities are usually different from those at low densities (Dreyfus, 1986). Additional physical phenomena appear, such as nonlinear effects or stimulated Raman or Brillouin scattering. These topics are dealt with in the standard books on lasers mentioned in Chapter 2. In addition to "intrinsic" effects, there are "extrinsic" effects that manifest themselves only at high power density levels:

(i) Fiber end preparation: As in the case of windows for high power transmission, there is a severe problem due to absorption at the two ends of the fiber. The end is usually cut and then polished and these two processes may leave some defects. Absorption of energy in these defects can cause local heating and damage. Good surface preparation of fiber tips is one of the crucial points for good transmission.

(ii) Internal defects: The same arguments apply for defects or impurities inside the fiber. Specially pure and defect-free fibers (power fibers) are therefore needed for high power densities (Schonborn *et al.*, 1986).

The following fibers were discussed in Chapter 4:

- Visible, near IR, and UV—fused silica fibers
- Near and mid-IR—fluoride and sapphire fibers
- Middle and far IR—crystalline fibers, hollow waveguides, and chalcogenide glasses

8.3.2 Components—Fiberoptic Tips

8.3.2.1 Modified Fiber Tips

In a regular fiber, the distal end is cut and polished. This end is often simply modified (Verdaasdonk and Borst, 1991a,b) as shown in Fig. 8.1 and Fig. 8.2.

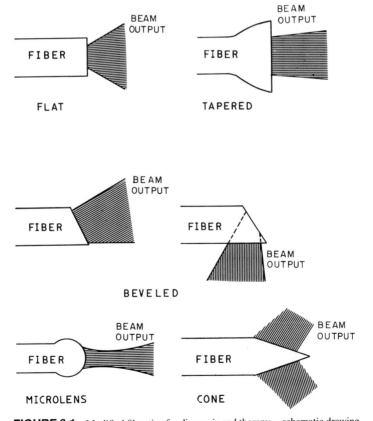

FIGURE 8.1 Modified fiber tips for diagnosis and therapy—schematic drawing.

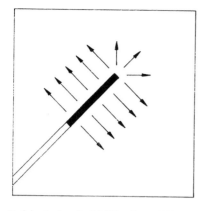 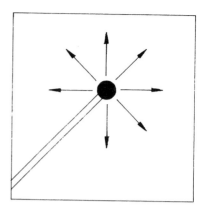

Cylindrical Diffusing Fiber   Spherical Diffusing Fiber

a

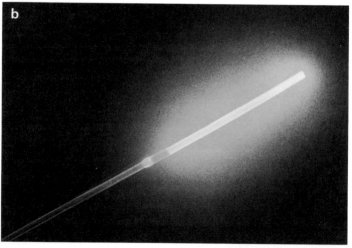

FIGURE 8.2 (a) Diffuser tips: cylindrical and spherical—schematic drawing. (b) Photograph of laser beam output from a diffuser tip. (Courtesy of PDT Systems.)

These modifications do change the output beam but they often cause some power loss.

- Tapered tips: A tapered end provides a less diverging beam (see Section 4.9), so that the power density at a distance of few millimeters from the tip is higher. This is important for laser surgery.
- Angled or cone-shaped tips: The beam output from an angled tip or a sharpened cone-shaped tip is diverted sideways. This may fill a need in some therapeutic methods, such as photodynamic therapy (PDT) or hyperthermia.

- Microlens tips: This tip can be designed to provide uniform illumination of tissue that is placed in front of the fiber.
- Ball-shaped tips: A lensed tip focuses the beam which then spreads. At the focal area, the laser power density is higher, allowing tissue ablation by increasing the energy density at the target.
- Diffuser: A tip whose surface has been roughened to increase scattering. Such tips, shaped like cylinders or spheres, spread the beam to the sides and are therefore termed diffusers. They have been used for HPD–PDT and for laser hyperthermia (Hasselgren *et al.*, 1990). The diffuser tips have been used in two ways:

(i) *Intraluminally*: The tip is placed within the lumen of a hollow organ, such as the esophagus or the colon, and provides illumination of the inner walls of the organ.

(ii) *Interstitially*: Several tips are inserted inside a large tumor and provide illumination of a sizable volume inside the tumor.

8.3.2.2 Attached Tips

There are other cases which require attaching special tips to the distal end of the fiber, as shown in Fig. 8.3.

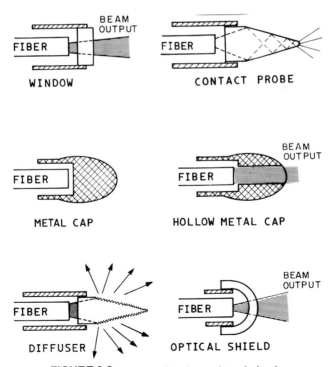

FIGURE 8.3 Attached fiber tips—schematic drawing.

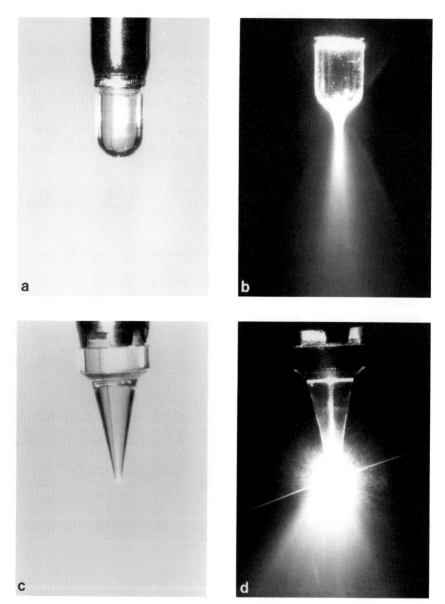

FIGURE 8.4 Photographs of attached sapphire tips and the beam shape exiting from them: (a) hemispherical tip of diameter 1.5 mm; (b) laser output in air; (c) conical (contact) tip; (c) laser output in water. (Courtesy of Surgical Laser Technology and R. Vedaasdonk.)

- Flat window: The window protects the fiber from debris or overheating. It is difficult to coat or deposit materials onto the polished faces of diamond or sapphire. If the window is made of material such as diamond, tissue and debris will not stick to it during surgery.
- Optical shield: A dome-shaped optical window which is used in contact with tissue (Cothren *et al.*, 1986).
- Contact probes: Cones made of sapphire or fused silica have also been used as contact probes, as shown in Fig. 8.4. The laser beam is concentrated at the tip of the cone and, upon contact with the tissue, the laser beam vaporizes it. It is probable that *in vivo* the laser energy causes carbonization of blood on the front surface of the tip. The energy is then absorbed in the carbonized layer and heats it to temperatures well above the ablation temperature of tissue. This may account for the efficiency of the contact probes (Daikuzono and Joffe, 1985).
- Metal tips: The metal cap and the hollow metal cap are heated by a laser beam (e.g., Nd:YAG or Ar laser); these have been used as "hot tips" for laser angioplasty (Hussein, 1986).
- Diffuser tips: These are attached tips whose surface is roughened. Their operation is similar to that discussed in the previous section.

Figure 8.5 shows photographs of a variety of modified and attached tips and the laser beam exiting from these tips.

8.3.3 Subsystem—The Fiberoptic Delivery Unit

The transmission of laser beams for therapy or diagnosis requires a fiberoptic delivery unit. This subsystem may also be considered as a simple laser catheter. A typical delivery unit is shown in Fig. 8.6 and consists of several components:

(i) Optical fiber: The fiber itself, with well-polished ends.
(ii) Connector: The input end of the fiber is often held in a connector that has been developed for communication purposes (e.g., one that is called an SMA connector).
(iii) Holder: The connector is inserted in a special holder that is fixed with respect to the laser beam.
(iv) Tip: A transparent window or tip at the output end of the fiber.
(v) A plastic jacket that strengthens and protects the fiber and may serve to hold the window or tip. The jacket also protects the fiber from sharp bending, which may damage the fiber or cause excessive optical losses.
(vi) A lens that focuses the laser beam into the input end of the fiber.

Either or both ends of the fiber may be overheated by the absorbed laser energy, especially when continuous-wave (CW) lasers are used. Cooling liquids or gases are often used, as shown in the figure, and the coolants exit through holes near the distal end of the delivery unit. Cooling gas or liquid must be used with

FIGURE 8.5 Photograph of various fiber tips. From right to left: 600-μm bare fiber; attached metal tip of diameter 2 mm; attached hollow metal tip of diameter 2.5 mm; attached hemispherical contact probe of diameter 2 mm; modified ball-shaped tip of diameter 1.5 mm. (Courtesy of R. Verdaasdonk.)

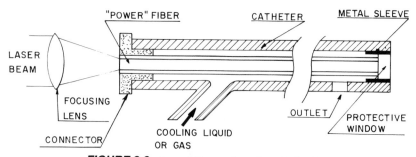

FIGURE 8.6 Laser delivery unit—schematic drawing.

caution in a body cavity because they can cause overdistention and perforation. This can be avoided by using a closed circulation of coolant inside the laser catheter.

Sometimes the whole assembly is provided by the manufacturer (as shown, for example in Fig. 8.7) and sometimes the delivery unit must be developed or modified by the user.

8.3.4 Systems—The Important Lasers

Each specific medical application is best performed by a particular laser operating at desired parameters, in conjunction with a suitable fiber optic system. The contents of the previous chapters (see Table 3.2 and Table 3.3) which discussed the different lasers and fibers are summarized in this section.

Biostimulation: Low-energy lasers may stimulate and accelerate biological processes such as the healing of wounds. HeNe lasers are often used for this application.

FIGURE 8.7   Photograph of an Nd:YAG laser–fiber system. (Courtesy of Laser Industries.)

Photochemical effects: Some chemical reactions are triggered by specific laser wavelengths. Gold vapor lasers, for example, emit the exact wavelength needed for HPD–PDT. Tunable dye lasers and the new AlGaInP lasers are also potentially useful for this application.

Photothermal effects: For heating larger volumes it is advantageous to use a laser whose radiation is not highly absorbed by tissue (so that the radiation penetrates deeply). Nd:YAG lasers are suitable for this purpose and have been used for coagulation and for hyperthermia. A laser beam whose wavelength is well absorbed by tissue, such as the excimer or the CO_2 laser, is effective for tissue vaporization and for laser surgery.

Photomechanical effects: When focused to a small spot on a target, pulsed dye or alexandrite lasers or Q-switched Nd:YAG lasers may generate plasma and shock waves. These effects are used for shattering of hard tissue.

8.4 THE DELIVERY OF HIGH-POWER LASER LIGHT THROUGH OPTICAL FIBERS—PRINCIPLES

The transmission of laser power through fibers, especially the transmission of short and energetic laser pulses, is a difficult problem, as discussed below. There are commercially available laser–fiber systems that give a partial solution to this problem and these systems are used clinically. However, the research scientist or physician who acquires the lasers and fibers may still face power transmission problems. A special section has thus been dedicated to some of the laser, fiber, and coupling problems.

8.4.1 Power Density

A fiberoptic delivery system basically needs to be capable of transmitting a sufficient energy supply to vaporize tissue. Consider a CO_2 laser beam of area 1 cm^2 which impinges on water at body temperature ($37°C$). The minimum energy needed to begin vaporization is roughly $5-10 \text{ J/cm}^2$ (see also Section 3.4.3.2 and Table 3.1). In many cases the laser is operated in the pulsed mode in order to reduce thermal damage (see Section 3.8.4).

Consider a few typical optical fibers of various diameters. In order to maintain an energy density, the thicker fibers must deliver more energy than the thinner fibers, as shown in Table 8.1.

Suppose that the laser operates in the pulsed mode. If we assume that the energy density is constant, $\epsilon/A = 10 \text{ J/cm}^2$, then for different fiber areas A we obtain different energies ϵ per pulse. This is shown in Table 8.2(I). We may also assume that the energy per pulse is constant $\epsilon = 100 \text{ mJ}$, and that the area of the fiber is constant, $A = 1 \text{ mm}^2$. The energy ϵ may be delivered in a short or a long pulse; the peak power p and the peak power density p/A are different, as shown in Table 8.2(II). It should be pointed out that for very short pulses, the power

TABLE 8.1 Pulsed Energy Delivery through Optical Fibers

Core diameter 0.2–0.6 mm
Core area 3×10^{-4}–3×10^{-3} cm^2
Pulse length 10–1000 ns
Energy per pulse 10–100 mJ
Energy density 10–100 J/cm^2
Pulse peak power 10^5–10^7 W
Power density 10^7–10^9 W/cm^2
Comments: Water vaporization 10 J/cm^2
          Ultrasound generation 10^7 W/cm^2
          Plasma (and shock wave) generation in water 10^9 W/cm^2
          Breakdown in air 10^{10} W/cm^2

TABLE 8.2 Pulsed Laser Transmission through Optical Fibers

I. Constant power density = 10 J/cm^2				
Core diameter (mm)	0.2	0.4	0.6	0.8
Core area (mm^2)	0.03	0.12	0.28	0.50
Pulse energy (mJ)	3	12	28	50
II. Constant pulse energy = 100 mJ; constant core area = 1 mm^2				
Pulse length	10 ns	100 ns	1 μs	1 ms
Peak power (W)	10^7	10^6	10^5	10^2
Power density (W/cm^2)	10^9	10^8	10^7	10^4

density that needs to be transmitted through the fiber is extremely high. This may cause laser-induced breakdown, as discussed below.

8.4.2 Laser Characteristics

Two characteristics of the laser beam are important for efficient coupling into a fiber:

1. Spatial uniformity: As explained earlier, the ideal laser beam has an intensity profile that is Gaussian. This beam shape is well suited for transmission in the optical fiber. Unfortunately, lasers are sometimes not well designed or aligned and the intensity distribution is far from Gaussian. In some pulsed lasers such as excimer or CO_2 TEA lasers there are "hot spots"—areas in the beam where the intensity is very high. When focused, the power density is particularly high at these points and may cause fiber damage. Manufacturers have been trying to get rid of hot spots and obtain a uniform beam.

2. Temporal behavior of pulsed lasers: Several lasers are potentially useful for surgical applications and emit very short pulses. The excimer lasers in the UV and the TEA CO_2 lasers in the IR emit energetic pulses (e.g., 0.01–0.1 J per pulse) whose length is less than 50 nsec. Quite often, these pulses consist of a very short

"spike" and a much longer "tail." Section 3.4.3 described the threshold for laser ablation and the dependence on the energy density. In order to vaporize tissue with a fiber whose core area is 1 mm^2, it is necessary to send a 100-mJ pulse of energy. Using a pulse length of 10 nsec, the power density is 10^9 W/cm^2 (see Table 8.2II). At these power densities, the fiber may break down (Taylor *et al.*, 1988). It has been shown that the damage threshold of the fibers depends on the pulse length. By lengthening the pulse length to 1 μsec, the power density is reduced to 10^7 W/cm^2, which may be transmitted without damage to the fiber.

8.4.3 Coupling Laser Power into a Fiber

In Section 4.8 we discussed power transmission through optical fibers. Let us now consider again some of the problems involved in coupling the laser beam by focusing it onto the input end of the fiber (Kar *et al.*, 1989; Kersten *et al.*, 1986). These problems have to be solved before one develops the fiberoptic delivery units that are discussed below.

(i) Acceptance angle: The fiber has a given numerical aperture, defining an acceptance angle. The focusing angle of the lens, which is determined by the numerical aperture of the lens, must be smaller than the acceptance angle of the fiber, as shown in Fig. 4.16a. Otherwise, a significant portion of the laser energy will be lost in the cladding or the outer plastic jacket. This may lead to the undesirable situation in which the jacket melts and the fiber is damaged.

(ii) Focal spot: As mentioned in Chapter 4, the lens may be adjusted to place the focal spot directly at the input end of the fiber (Fig. 4.16b). This situation, however, is undesirable because the power density at the focal spot is extremely high and the input end of the fiber is likely to be damaged. In order to avoid this, the fiber end may be moved away from the focal spot (Fig. 4.16c). The lens is used to form a spot on the input end with an area roughly equal to or slightly smaller than the core of the fiber. This reduces the possibility of damage.

(iii) Alignment: The laser beam must be aligned with the fiber axis (Fig. 4.16d), otherwise excessive losses will occur. Often the lens is attached to the laser which focuses the laser beam onto a fixed point in space. The input end of the fiber is held in a mechanical attachment which facilitates positioning the fiber core near the focused beam.

8.5 INTEGRATED LASER—FIBER SYSTEMS AND THEIR APPLICATIONS—PRINCIPLES

8.5.1 Integrated Systems

(i) Laser–fiber: This system comprises a laser source and an optical fiber delivery unit. It is the basic ingredient of laser catheters and endoscopes.

(ii) Laser catheter: A compound catheter may include several open channels. The power fiber of a laser–fiber system may be inserted through one of the chan-

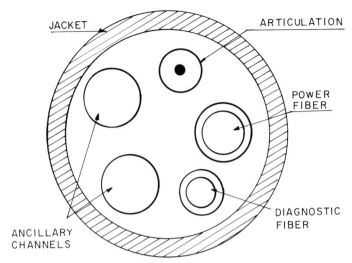

FIGURE 8.8 Cross section of a laser catheter.

nels and the laser power is used for therapy. The optical fiber of a fiberoptic sensor may be inserted through another channel and used for diagnosis. Other channels may be used for introducing guide wires (see Section 9.2.2) or for injection of fluids. In some catheters the distal tip may be flexed, using some angulation mechanism (see Section 6.2.1.2). A cross section of the laser catheter is quite similar to the cross section of a laser endoscope (shown in Fig. 8.12) but without the imaging and the illumination bundles. A cross section of a catheter with several ancillary channels, with power and diagnostic fibers, and with an articulation mechanism is shown in Fig. 8.8; photographs of a laser catheter are shown in Fig. 8.9.

(iii) Laser endoscope: In this system, the laser–fiber and the fiberoptic sensor systems are incorporated in an endoscope system. In addition to therapy and diagnosis, this integrated system provides imaging. A schematic drawing of an ordinary rigid endoscope with fiberoptic illumination is shown in Fig. 8.10. The corresponding laser endoscope is shown schematically in Fig. 8.11. A cross section of a fiberoptic laser endoscope is shown in Fig. 8.12.

8.5.2 Diagnosis of Disease

This section looks at the optical system used for identifying diseased tissue and for performing some diagnostics.

(i) Blood clot: Intraluminal blood clots can be observed by angioscopy because their color and consistency are different from those of normal blood. Blood clots blocking the blood flow can be diagnosed by the methods discussed in Section 7.5.

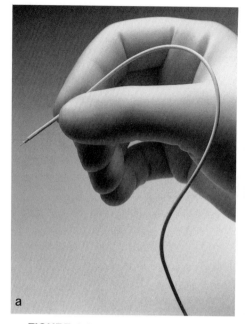

FIGURE 8.9 Photograph of a laser catheter: (a) general view; (b) distal end. (Courtesy of Coherent.)

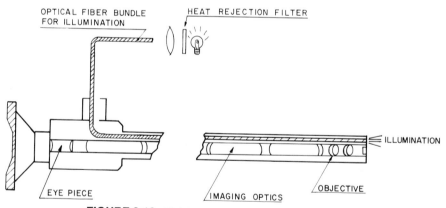

FIGURE 8.10 Rigid endoscope with light guide illumination.

    (ii) Malignant tumors: In addition to direct observation by the endoscopist, tumors can be distinguished from healthy tissue by measuring exogenous or endogenous laser-induced fluorescence (LIF). This is particularly important for tumors that are too small to be detected by nonoptical methods such as ordinary x-ray imaging or CT.

    (iii) Atherosclerotic plaque: Although this tissue is generally soft, it may be

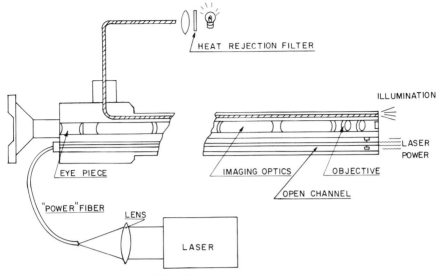

FIGURE 8.11 Rigid laser endoscope.

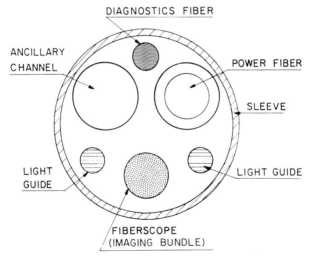

FIGURE 8.12 Cross section of a fiberoptic laser endoscope—schematic drawing.

hard if calcified. Compared to normal tissue, the plaque may have a characteristic luminescence (autofluorescence). Selective uptake of dyes is also used to locate plaque.

(iv) Hard tissue: This includes bone fragments that must be removed from joints. These fragments have different colors and are usually anatomically obvious to the endoscopist.

(v) Stones: Urinary or biliary stones are most often detected by x-rays but can be seen and treated endoscopically. In laser lithotripsy, stones are exposed to intense laser pulses. A plasma is generated by these pulses under water, and the resulting shock waves shatter the stones. The plasma gives rise to a characteristic light emission which may be measured through the fiber and used for identifying the stones. Incidentally, the shock wave is also transmitted to the proximal end of the fiber. It may be detected acoustically, thus indicating that the stone is being shattered.

(vi) Dental tissue: There is a marked difference in the fluorescence spectra emitted from carious and noncarious regions of the tooth. This may be used as a diagnostic tool during therapy as well as for the early detection of tooth decay.

8.5.3 Laser Therapy

Laser therapeutic methods may be classified in the following way (see Section 3.6):

- (i) Laser heating at very low energies—biostimulation.
- (ii) Laser heating at low energies—coagulation, welding, and hyperthermia.
  Coagulation: Heating to 60–70° C causes contraction of tissue around bleeding vessels, which are then squeezed shut.
  Welding: Laser energy can perform anastomosis and welding of tissues at typical temperatures of 60–70° C.
  Hyperthermia: Laser heating at typical temperatures of 42.5–43° C is used for hyperthermia cancer treatment.
- (iii) Laser heating at medium energies—vaporization. Sufficient laser heating can vaporize tissue, drill holes, and cut and ablate tissue.
- (iv) High-energy laser pulses—lithotripsy and photoablation.
  Lithotripsy: Shattering of urinary stones by shock waves generated by powerful laser pulses.
  Photoablation: Short, energetic laser pulses vaporize tissue with little thermal damage.
- (v) Nonthermal effects—therapy and tissue removal.
  Photochemical methods: Laser energy is absorbed either by the tissue itself or by dyes or drugs introduced into the tissue. The laser energy triggers therapeutic photochemical reactions.
  Photodisruption: Excimer laser removes tissue with no signs of thermal damage. There are claims that this technique is photodisruptive and not thermal in character.

8.5.4 Monitoring and Control; Dosimetry

The integrated system may consist of a laser catheter or a laser endoscope. It may or may not incorporate fiberoptic diagnosis. A laser catheter is inserted into the body and its position can be monitored by nonoptical systems such as x-ray systems. In the case of laser catheters, for laser angioplasty, the effects of the delivered laser beam can also be monitored by x-ray fluoroscopy and angiographic contrast methods. In more sophisticated systems, an additional fiberoptic subsystem monitors and controls the actions of the therapeutic laser beam. In the case of laser endoscopes, the monitoring and control are done with the help of the imaging component of the endoscope, which can also incorporate fiberoptic subsystems for diagnostics.

Laser therapy is dependent in many cases on the energy absorbed in the tissue. It is therefore important to determine the fluence (also called dose) that is delivered through the fiber. This problem is sometimes referred to as dosimetry. Several devices were tried in a few clinical applications, but the general problem of *in situ* dosimetry has not yet been solved satisfactorily.

8.5.5 Integrated Laser–Fiber System—Summary

Each of the integrated laser–fiber systems may be divided into systems, subsystems, and components. This division is shown in Table 8.3. A photograph of a laser–endoscope integrated system is shown in Fig. 8.13.

TABLE 8.3 Integrated Laser–Fiber Systems

	Description	Structure	Applications
Component	Fiber	Fiber	Light transmission
	Nonordered bundle	Array of optical fibers	Illumination
	Ordered bundle	Ordered array	Image transmission
Subsystem	Laser	Laser	Source of light
	Fiberscope	Ordered and nonordered bundles	Image transmission
	F/O[a] delivery unit	Input coupler + optical fiber + output window	Delivery of laser light
System	Endoscope	Fiberscope + auxiliary channels	Imaging; insertion of endoscopic devices
	F/O sensor	Laser + fiber + optical processing	Sensing
Integrated system	Laser fiber	Laser + F/O delivery unit	Delivering laser power for diagnosis or therapy
	Laser catheter	Laser + fiber in a catheter	Performing laser diagnosis and therapy inside the body
	Laser endoscope	Laser + fiber in an endoscope	Performing imaging, diagnosis, and therapy

[a]F/O = fiberoptic.

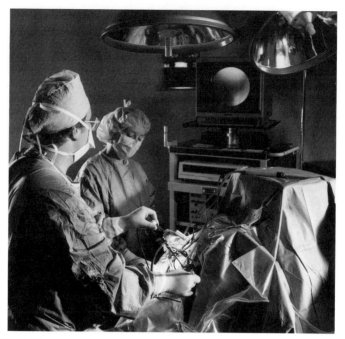

FIGURE 8.13   Photograph of an integrated laser endoscope system. (Courtesy of Coherent.)

8.6 OPERATION OF PRACTICAL FIBEROPTIC LASER SYSTEMS—PRINCIPLES

This section describes some of the systems currently being used either clinically or in preclinical studies. The possible selection of lasers, suitable power fibers, and suitable tips is shown in Table 8.4.

TABLE 8.4   **Laser and Fiber Systems**

Laser (general)	Wavelength (nm)	Delivery system	Distal end tips
Excimer	193, 249	Articulated arm	?
	308, 350	SiO_2 fiber	SiO_2
Ar ion	488, 514	SiO_2 fiber	Sapphire
Nd:YAG	1060	SiO_2 fiber	Sapphire
Er:YAG	2.9 μm	ZrF_4 fiber?	Sapphire
			Diamond
CO_2	10.6 μm	Halide fibers	Diamond
		Hollow guides	ZnSe

8.6.1 Argon Ion CW Laser Fiberoptic Systems—Heating

The laser: The Ar ion laser emits blue–green light mainly at 488 and 514 nm. The emission is continuous, with power levels of 10–20 W.

The fiber: Ar laser light can be transmitted by good-quality, silicate glass fibers. Fused silica (quartz) fibers can handle higher power and are preferable for this application.

Laser–fiber coupling: Coupling is straightforward, based on good-quality optical lenses which transmit visible light.

Fiber tip: The distal tip can either be bare or have a specific attachment; for example, the hot tip is metallic. When the tip is heated by the laser beam, it is able to heat and melt tissue.

The system: The fiber can be inserted into a catheter to form a laser catheter. Such a catheter can incorporate another thin fiber for diagnostic purposes. The fiber may also be inserted into a thin endoscope to form a laser endoscope.

Typical applications: The system which is based on an Ar laser and a glass fiber with a bare tip was one of the first systems used in laser angioplasty. The system based on the Ar laser and on fused silica fibers with a hot tip at the end is also used for laser angioplasty. Although most of the systems are laser catheters, laser endoscopes are being considered as well.

8.6.2 Nd:YAG CW Laser Fiberoptic System—Heating

Laser: The Nd:YAG laser emits near-infrared radiation at 1.06 μm. This radiation penetrates a few millimeters into tissue and thus heats the tissue. It can also vaporize and remove tissue. Most Nd:YAG lasers used in medicine emit CW radiation of a few tens of watts.

Fiber: The near-infrared radiation of such high intensity is transmitted by fused silica fibers.

Laser–fiber coupling: Good-quality optical lenses are used for straightforward coupling.

Fiber tip: For some applications, bare fibers are used. Other applications use conical sapphire "contact" tips. These tips facilitate cutting with the Nd:YAG lasers (see Section 8.3.2). At lower laser powers, sapphire tips have also been used as diffusers for laser hyperthermia. Metal tips can also be used with the Nd:YAG lasers.

System: Nd:YAG lasers have been used with fused silica fibers that are inserted into catheters or into endoscopes (Frank, 1986). In some applications tens of watts are continuously transmitted through the fiber and it must be cooled. This is achieved through a subsystem that provides pressurized cooling gases which pass along the fiber.

Typical applications: If the distal tip of the fiber is bare, the laser radiation can penetrate deep into the tissue. Endoscopes based on this laser are widely used for blood coagulation in gastroenterology and for bladder tumor treatment in urology. Laser catheters based on fused silica fibers with sapphire tips have been used for tissue removal. In addition to benign prostatic hypertrophy and prostrate cancer, they have also been used in the bronchoscopic ablation of bronchial tumors.

Incidentally, one laser–fiber system is somewhat different. Its source is an Nd : YAG laser and the waveguide is a jet of water that flows in a medical catheter (Sander *et al.*, 1988; Gregory and Anderson, 1990). It has been suggested for clinical applications in laser angioplasty or for coagulation in gastroenterology. From the historical perspective this system reminds one of the water jet system which was used 100 years ago by Tyndall in the early demonstration of light guiding by internal reflections (see Section 4.2).

8.6.3 Excimer Pulsed Laser Fiberoptic Systems—Ablation

Several excimer lasers are based on different gases. Some emit at 308 nm and some emit at shorter wavelengths (see Table 2.3). In all cases, the lasers emit rather short pulses (of 10–100 nsec) with relatively high pulse energy.

Fibers: Present optical fibers cannot transmit the shorter wavelengths ($\lambda <$ 300 nm) emitted by excimer lasers. Fused silica fibers are used for longer wavelengths, such as 308 nm. It is more difficult to transmit short energetic pulses than to transmit CW radiation. Because the fibers may be damaged by the high power density, it has been found that it is easier to transmit 100-nsec pulses of lower-peak power than 10-nsec pulses of higher-peak power and longer-pulse excimer lasers have been built for this purpose. The first fibers used in laser angioplasty were fused silica fibers of diameter greater than 1 mm. Although these fibers did transmit the laser energy, they were much too stiff to be practical. It was suggested that the single fiber be replaced by a bundle of thin fibers. A special bundle was fabricated so that the fibers formed a full circle at the input end and a ring at the output end (see Fig. 5.1). This "ring catheter" is hollow so that it can be passed over a guide wire. The excimer laser beam is focused on the input end. Near the output end, the beam emitted from the ring of fibers is shaped like an open tube. This has been found useful for "drilling" large holes in atherosclerotic plaque.

Laser–fiber coupling: Optical elements used to focus the beam onto the fiber, for 308 nm radiation, are also made of fused silica. The beam emitted from the excimer laser may be nonuniform and contain hot spots, which can cause damage when the beam is focused on the proximal end of the fiber. The coupling of the excimer laser beam into the fiber is therefore a complex procedure.

Fiber tip: In most applications, the excimer laser beam interacts directly with tissue and there is no need for special tips. Nevertheless, ball-shaped tips may help prevent mechanical perforation of the blood vessel wall in laser angioplasty.

System: Both laser endoscopes and laser catheters based on excimer lasers have been developed (Taylor *et al.*, 1988; Kubo *et al.*, 1989).

Typical applications: The pulsed laser radiation cuts well without damaging the neighboring tissue. It is mostly used in laser surgery, laser angioplasty, or laser diskectomy.

8.6.4 CO_2 Laser–Fiber System—Cutting

Laser: Several types of CO_2 lasers are used for medical applications. Common CO_2 lasers are based on a gas flow system. They emit CW radiation with power levels of a few tens of watts, and some of them are capable of emitting a train of long pulses. Lasers such as the waveguide CO_2 laser are small and portable. They emit radiation with powers of 10–30 W. TEA CO_2 lasers have a mechanical and optical structure similar to that of excimer lasers, but they use CO_2 gas as a medium. They also emit very short (10–100 ns) and energetic (several joules) pulses, but the emission is in the IR (10.6 μm). Since they cause very little damage to surrounding tissue, TEA lasers are potentially useful for surgical applications.

Fiber: Very few fibers can transmit the 10.6-μm radiation of the CO_2 laser and even fewer can handle high power at this wavelength. At present, only crystalline fibers made of metal halides or hollow waveguides are successful.

Coupling: Infrared transmitting lenses (e.g., made of ZnSe) are used to couple the laser radiation to the fiber. In the case of CW lasers, the beam often has a Gaussian shape and it is not too difficult to couple the beam into the fiber.

System: Few experiments have been performed with laser catheters and laser endoscopes because of the lack of suitable fibers. In principle, this system is similar to the systems already described (Gal and Katzir, 1987; Ishiwatari *et al.*, 1986).

Typical applications: CO_2 laser radiation is most useful for cutting, for removal of tissue, and for laser surgery. Some clinical work has utilized the CO_2 laser and hollow waveguides to remove tissue inside joints by way of an arthroscope. If the fiberoptic problem can be solved, the fiberoptic systems will be widely used for endoscopic laser surgery.

8.6.5 Pulsed Dye and Nd : YAG Laser Fiberoptic Systems—Lithotripsy

Various lasers and power fibers have been used for laser lithotripsy by many groups. Some of the conclusions obtained by the different researchers are as follows:

- CW lasers: Nd : YAG, Ar, and CO_2 lasers have been tried. The thermal interaction between these lasers and the stones has not been effective for stone destruction.
- Short-pulse lasers: One of the first successful methods for fragmenting stones is based on Q-switched Nd : YAG lasers which operate at 10–20 Hz. They emit short (10 nsec) and energetic (1 J) pulses which can be trans-

mitted by thin (0.6 mm) fused silica fibers. When the distal tip of the fiber is brought to the vicinity of the stone, plasma is formed near the surface and a shock wave is generated. This shock wave gradually shatters the stone. The major problems with this method are how to couple such high peak power pulses into the fiber without destroying the proximal end and how to prevent the distal end from being destroyed during the stone shattering.

Excimer laser pulses have also been sent through fibers and tested for stone removal. The action of the excimer laser is based on material ablation and not on shock wave formation. Because the ablation process is much less effective, the excimer lasers are not widely used for this purpose.

- Intermediate-length pulses: Although several lasers have been tried, the most effective is the flashlamp-pumped tunable dye laser. The laser energy is coupled into a thin (0.2 mm) fused silica fiber and impinges on stones immersed in liquids. Laser effects have been studied as a function of the laser wavelength, pulse duration, and power density. The laser pulse produces plasma near the surface of the stone which shatters the stone by producing a propagated shock wave. Optimal conditions were obtained for a wavelength of 504 nm, which is highly absorbed by the yellow or black pigments in the stone but not by the hemoglobin of the surrounding tissues. At this wavelength, a pulse duration of 1 μsec is chosen and the threshold for stone shattering is 20 mJ per pulse. Another laser under study for stone fragmentation that has shown very promising results is the Alexandrite laser system.

8.7 FLOWCHART DIAGRAMS—ADVANCES

Many problems in science and engineering involve a number of complicated decision-making processes. One of the commonly used methods for logical decision making is based on the concept of flowchart diagrams. The total process is divided into a series of steps. The first and the last step are Start and Stop. The progress from step to step is shown schematically by arrows. Simple steps (i.e., actions) are shown schematically inside rectangular brackets. More complicated steps, which involve decision making, are shown in rhombus brackets. If the decision is "yes," the arrows continue in one direction; if the decision is "no," another direction is chosen. The flow diagram thus helps in visualizing a complicated series of steps.

The integrated laser–fiber systems discussed in this chapter may be rather complex. The application of these systems for procedures such as laser angioplasty or HPD–PDT is fairly complicated too. It may therefore be of benefit to describe the procedures (and the use of the integrated systems) by flowchart diagrams.

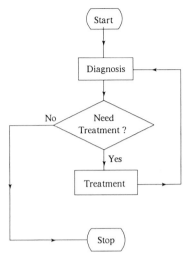

FIGURE 8.14 Flowchart diagram for the use of the integrated systems for diagnosis and treatment.

The hypothetical example shown in Fig. 8.14 will clarify the usefulness of flowchart diagrams. The first step is Start. The next step is a Diagnosis step and may involve endoscopic imaging or fiberoptic sensing. This leads to a decision-making step: Need Treatment? If a disease is not found, this is the end of the procedure and one moves to Stop. If a disease has been found, the course of therapy is determined and Treatment is chosen. At this stage, the suitable laser–fiber system and the mode of operation are decided upon. After the treatment, Diagnosis is carried out again. The next step is again the decision making: Need Treatment? If it is not necessary to carry out any further treatment, the whole process goes to Stop. Otherwise, one goes back to the decision-making step and the process continues until the treatment has accomplished its goals.

8.8 COMPLICATIONS IN THE USE OF LASER FIBEROPTIC SYSTEMS—PRINCIPLES

This section mentions several problems that are common to some of the laser fiberoptic systems mentioned above. Many of these problems are likely to be solved in the future.

8.8.1 Mechanical Problems in the Systems

(i) Optical fibers tend to break upon repeated bending. This limits the operation of ultrathin endoscopes and may also be a problem with thin power fibers.

(ii) Thick optical fibers that are often used for power transmission are stiff and difficult to use. These fibers may also break upon sharp bending and may perforate the catheter or the endoscope in which they are located.

(iii) The mechanical devices that connect fibers to holders, or that attach windows or tips to the ends of fibers, are still bulky and cumbersome to use.

(iv) Thin and ultrathin catheters and endoscopes cannot be easily inserted and guided inside the body but must be inserted through larger catheters or over a guide wire.

(v) There is no satisfactory method for angulation of the distal tips of these devices. This is important, for example, in endoscopic laser surgery, where the output beam should be directed only toward a given target tissue.

(vi) Present systems require gases or liquids to cool the power fibers. A special subsystem is needed to circulate these fluids.

(vii) The interaction between the laser beam and tissue generates gaseous products which have to be evacuated.

(viii) Sterilization of the ancillary channels in thin endoscopes is difficult.

8.8.2 Optical Problems in the Systems

(i) Focusing a high-power laser beam into a thin optical fiber presents a problem.

(ii) Dosimetry: it is difficult to asses how much laser energy has reached a targeted tissue and how much has actually been absorbed in the tissue.

(iii) At high power densities there are deleterious effects at the output face of the fiber or at the interface between the output face and an attached tip. The input or the output face is often damaged or melted.

(iv) An optical feedback and control mechanism that will prevent this damage has not yet been developed.

(v) In endoscopy, there are some fundamental problems which have not been solved, such as the two-dimensional nature of the image and the difficulty in determining the size of an observed object. These problems are particularly important in the case of ultrathin endoscopes.

8.8.3 Effects Generated by the Systems

In all the therapeutic and surgical systems mentioned earlier, a laser beam is sent through a power fiber and interacts with internal organs. If the average power delivered is too high, the laser beam may cause thermal damage to neighboring tissues, such as undesirable blood coagulation and thrombosis.

If the laser beam is directed in a wrong direction, it may vaporize inappropriate areas. This may happen with both bare fibers and fibers with attached tips. For example, in laser angioplasty, a misdirected laser beam may perforate a blood vessel wall.

Interaction of the laser beam with soft or hard tissue or with blood may cause debris. Concern has been expressed that particulate or gaseous debris may present toxicity problems. In particular, it was thought that debris may cause embolization in laser angioplasty. Recent studies indicate that these problems are not severe.

In most surgical systems, gases or liquids are used to cool the power fibers and these may cause embolisms.

8.8.4 Safety

In the case of laser–fiber systems, laser energy is transmitted through thin, flexible fibers that are inserted into catheters or endoscopes. There is some danger that the physician or the personnel may be exposed to direct laser energy emerging from the distal tip of the fiber. This may happen if the fiber slips out of the catheter or endoscope or if it is taken out to clean debris and the operator forgets to turn off the laser. Although it is true that the beam emerging from the tip is diverging, it may still exceed the maximum permissible exposure (MPE) of the eye. Physicians, attendant staff, and patients must wear suitable protective eyewear.

Special care should be taken when lasers are used in integrated systems such as the laser endoscope or laser microscope. In these systems, a laser beam is transmitted onto tissue through a lens or an optical fiber and the physician looks at the same tissue through an endoscope or a microscope. In the case of laser surgery or laser therapy (e.g., PDT), the power transmitted through the fiber may be of the orders of several watts. The laser beam may be inadvertently reflected back into the physician's eye and cause damage, unless protective eyewear is used.

References

Apfelberg, D. B. (1987). *Evaluation and Installation of Surgical Laser Systems*. New York: Springer-Verlag.

Artjushenko, V. G., Dianov, E. M., Konov, V. I., Nikiforov, S. I., Prokhorov, A. M., Silinok, A. S., and Shcherbakov, I. A. (1989). Promising laser fiber systems for surgery. *Proc. SPIE* **1067**, 233–241.

Cothren, R. M., Kittrell, C., Hayes, G. B., Willett, R. L., Sacks, B., Malk, E. G., Ehmsen, R. J., Bott-Silverman, C., Kramer, J. R., and Feld, M. S. (1986). Controlled light delivery for laser angio-surgery. *IEEE J. Quantum Electron.* **QE-22**, 4–7.

Daikuzono, N., and Joffe, S. N. (1985). Artificial sapphire probe for contact photocoagulation and tissue vaporization with the Nd:YAG laser. *Med. Instrum.* **19**, 173–178.

Dreyfus, M. G. (1986). Glass requirements in medical fiber optics. *Adv. Ceram. Mater.* **1**, 28–32.

Frank, F. (1986). Multidisciplinary use of the NdYAG laser. *Proc. SPIE* **658**, 22–25.

Gal, D., and Katzir, A. (1987). Silver halide optical fibers for medical applications. *IEEE J. Quantum Electron.* **QE-23**, 1827–1835.

Gregory, K. W., and Anderson, R. R. (1990). Liquid core light guide for laser augioplasty. *IEEE J. Quant. Elect.* **12**.

Hasselgren, L., Galt, S., and Hard, S. (1990). Diffusive optical fiber ends for photodynamic therapy: manufacture and analysis. *Appl. Opt.* **29**, 4481–4488.

Hussein, H. (1986). A novel fiberoptic laserprobe for treatment of occlusive vessel disease. *Proc. SPIE* **605**, 59–66.

Ishiwatari, H., Ikedo, M., and Tateishi, F. (1986). An optical cable for a CO_2 laser scalpel. *J. Lightwave Tech.* **LT4**, 1273–1279.

Kar, H., Helfmann, J., Dorschel, K., Muller, G., Muller, O., Ringelhan, H., and Schaldach, B. (1989). Optimization of the coupling of excimer laser radiation into Q-Q fibers ranging from 200–600 micrometer diameter. *Proc. SPIE* **1067**, 223–232.

Kersten, R. T., and Kobayashi, N. (1986). Laser beam delivery system for medical applications. *Proc. SPIE* **658**, 28–31.

Kubo, U., Hashishin, Y., and Okada, K. (1989). UV beam guide for medical excimer lasers. *Proc. SPIE* **1067**, 204–210.

Sander, R., Poesl, H., Frank, F., Meister, P., Strobel, M., and Spuhler, A. (1988). An Nd:YAG laser with a water guided laser beam—a new transmission system. *Gastrointest. Endosc.* **34**, 336–338.

Schonborn, K. H., Kobayashi, N., and Kersten, R. T. (1986). High power laser beam delivery systems in surgery: the technical aspect. *Proc. SPIE* **658**, 32–35.

Taylor, R. S., Leopold, K. E., Brimacombe, R. K., and Mihailov, S. (1988). Dependence of the damage and transmission properties of fused silica fibers on the excimer laser wavelength. *Appl. Opt.* **27**, 3124–3134.

Verdaasdonk, R. M., and Borst, C. (1991a). Ray tracing of optically modified fiber tips 1. Spherical probes. *Appl. Opt.* **30**, 2159–2171.

Verdaasdonk, R. M., and Borst, C. (1991b). Ray tracing of optically modified fiber tips. 2. Laser scalpels. *Appl. Opt.* **30**, 2172–2177.

9

Clinical Applications of Fiberoptic Laser Systems

9.1 INTRODUCTION

Previous chapters did not detail the usefulness of the integrated laser–fiber system in the various medical disciplines. This chapter explains the laser and fiberoptic methods for each discipline. It demonstrates their potential applications and provides the physician with sufficient background to start reading the latest literature. The various medical disciplines are arranged in alphabetic order. References at the end of the chapter and the bibliography give more details on the specific diagnostic or therapeutic applications. The various diagnostic and therapeutic modalities are summarized and are sometimes illustrated by flowcharts. The clinical applications of lasers are given in Table 8.1.

Each section consists of basic information, followed by the principles of the particular diagnostic and therapeutic methods, and illustrated with clinical results. Emphasis is placed on the systems and methods which have already been put into practice or have been tried in preclinical and clinical experiments. Section 9.2 is somewhat longer, not because cardiology is more important than the other disciplines but primarily because of the technical complexities involved in this field. Also, this is the first section, and many of the technical aspects mentioned in Section 9.2 have relevance to the other sections.

Several medical disciplines such as dermatology and dentistry have not been discussed in detail in the ensuing sections. Although lasers play a major role in dermatology (Andre, 1990), there has been little use of optical fibers. On the other hand, fiberoptic laser systems are potentially useful in dentistry. The interaction between lasers and dental tissue was discussed in the 1960s (Stern and Sognnaes, 1964) and the use of ultrathin endoscopes for viewing root canals was tried in the

1980s (Marshall *et al.*, 1981). In Section 7.4.2 we mentioned that laser–fiber systems may be used to diagnose or treat root caries. However, lasers have just started to penetrate dentistry and oral surgery (Willenberg 1989; Midda and Renton-Harper 1991) and it will be some time before fiberoptic systems are used in dentistry (see Table 9.1).

9.2 FIBEROPTIC LASER SYSTEMS IN CARDIOVASCULAR DISEASE

9.2.1 Introduction

During the past few decades, cardiology has become a prominent specialty in medicine. Cardiovascular disease is one of the principal causes of death worldwide. Major surgery and other invasive modalities are often required to diagnose and treat the disease.

The heart, coronary arteries, and peripheral arteries all constitute a rather complex system of tubes, pumps, and valves which lends itself to the use of fiberoptic investigation and treatment. Fiberoptic imaging can be used to identify a diseased area. Laser–fiber systems can be useful to diagnose and treat cardiovascular disease. Laser angioplasty has already been mentioned. Some of the developments in this area are discussed in recent books (Abela 1990; Litvack 1992; Sanborn, 1989; White and Grundfest 1989) and review articles (Cragg *et al.*, 1989; Isner and Clark, 1984; Isner *et al.*, 1987; Michaels 1990; Waller 1989).

9.2.2 Endoscopic Laser Systems in Cardiology—Fundamentals

9.2.2.1 Fiberoptic Laser Systems

The fiberoptic and laser methods used in cardiology are as follows:

(i) *Guidance*: Optical fibers used for therapeutic or diagnostic purposes are placed near a blockage in an artery with either a catheter or endoscope. The catheter is used by techniques that have been developed for regular angiography. A long, flexible guide wire, inserted into a peripheral artery in the groin or arm, is advanced in the arterial system toward the coronary arteries. The physician twists and bends the guide wire externally. A torque is transmitted to the distal tip of the wire, giving control of the distal tip position. Its position is monitored by x-ray fluoroscopy. When the guide wire is in place, the physician slides a thin catheter over it and into the coronary artery. The guide wire is then pulled out. With the catheter now in position, the physician often injects a radiopaque liquid (also called contrast medium) into the artery. This liquid is opaque in the sense that it highly absorbs x-rays and is clearly seen in fluoroscopic angiography. A thin optical fiber can be inserted into the same catheter and pushed all the way into the artery until its distal tip is brought into contact with the atherosclerotic blockage, as shown in Fig. 9.1.

TABLE 9.1 **Clinical Applications of Lasers**

Specialty	Laser	Applications
Cardiology	Excimer Ar Dye Nd:YAG	Laser angioplasty; endarterectomy
Dentistry	CO_2 Nd:YAG	Soft tissue surgery Caries removal
Dermatology	Ar Dye	Port wine stains and strawberry marks Varicose vein and tattoo excision; spider nevi
Gastroenterology	Ar Nd:YAG Dye	Esophageal, gastric, and colorectal polyps and carcinoma; bleeding lesions (ulcer, gastritis) Esophageal, gastric colorectal, and biliary polyps and carcinoma; bleeding varices, ulcer, gastritis, and angiomata; tumor palliation and cure Biliary stones
General Surgery	CO_2 Nd:YAG	General cutting tool; welding Cholecystectomy
Gynecology	Ar Nd:YAG CO_2 Ar Nd:YAG CO_2	Laparoscopic surgery: endometriosis, tubal surgery, adhesiolysis Menorrhagia Fallopian tube reconstruction; herpes; infertility cervical conization; valvular carcinoma
Neurosurgery	CO_2	Tumor excision; intracerebral surgery; meningioma and glioma excision
Oncology	Dye Au CO_2 Nd:YAG	Photodynamic therapy of tumors Tumor debulking Hyperthermia; thermotherapy of tumors
Otolaryngology	Ar Nd:YAG CO_2	Bleeding lesions; subglottic hemangioma Tracheal webs and hemangioma; hemostasis Polyp excision; tracheal webs and stenosis; papillomas; tracheobronchial carcinoma
Ophthalmology	Excimer Ar ion Nd:YAG	Corneal surgery Retinal detachment; iridectomy; proliferative retinopathy; glaucoma; senile macular degradation Posterior capsulotomy; iridectomy; vitreous bands
Orthopedics	CO_2 Ho:YAG	Arthroscopic surgery; bone tumor excision
Urology	Ar ion Dye Nd:YAG Nd:YAG CO_2	Urethral stricture; bladder hemorrhage; bladder tumor excision Laser lithotripsy (kidney stones) Bladder bleeding; bladder tumor therapy Renal resection; penile carcinoma; circumcision: urethral stricture

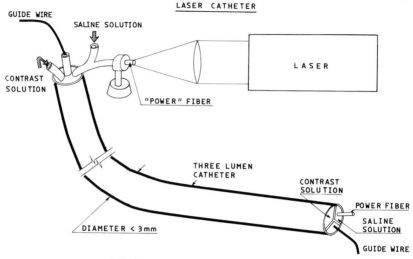

FIGURE 9.1 Laser catheter for cardiology.

(ii) *Imaging*: The development of thin and ultrathin endoscopes paved the way for fiberoptic imaging inside blood vessels. These endoscopes, however, are not as rigid as the guide wire. A torque cannot be applied to the proximal end of the endoscope and transmitted to the distal tip. At present, thin endoscopes can be guided through blood vessels by two methods:

- Guide catheter: The method mentioned in (i) is the simplest for inserting a guiding catheter into a desired artery. The ultrathin endoscope is inserted through this catheter, which is then pulled out, leaving the endoscope in place.
- Guide wire: The wire is inserted first and the endoscope slides over it.

The fiberoptic imaging is performed with regular white light. Image enhancement techniques or fluorescent imaging (see Section 6.5.1) can also be used. In the near future, this imaging method will probably be used to complement the more widely used imaging methods, such as x-ray fluoroscopic angiography or magnetic resonance imaging (MRI). In the far future, fiberoptic imaging may be an independent method, serving as one of the important tools of the cardiologist.

(iii) *Diagnosis*: Optical fiber sensors can be inserted into blood vessels or the heart via thin catheters or thin endoscopes. All the diagnostic methods mentioned in Chapter 7 are applicable in cardiology and most of them have already been tried clinically.

(iv) *Therapy*: Laser angioplasty has been performed to recanalize blockages in the coronary or peripheral arteries in thousands of patients. Lasers can also be used for endarterectomy (the removal of plaque) or for tissue welding in the cardiovascular system.

(v) *Other endoscopic techniques*: Thin endoscopes can be used for many of the therapeutic methods mentioned in Section 6.2, such as injection of drugs or dyes into atherosclerotic blockages or insertion of ultrasonic imaging devices to measure the thickness of the blood vessel wall. These are still under investigation. A complete laser endoscopic system is shown in Fig. 9.2.

9.2.2.2 Mechanical Devices

Several mechanical devices were proposed for atherectomy, the excision of atheroma inside blood vessels (Forrester *et al.*, 1991). Each of these devices is based on a atherectomy catheter which has to be guided and positioned.

(i) *Directional atherectomy* devices (Ellis *et al.*, 1991) are based on a probe with a rotary cutter of diameter 1.5–2.5 mm at its end. This atherectomy catheter is introduced through a guiding catheter and its tip is placed near the stenosis. The atheroma is excised using the rotary cutter and collected in a nose cone. It is removed from the artery when the cutter is withdrawn.

(ii) *Rotational atherectomy* devices (Fourrier *et al.*, 1989) are based on a rotating abrasive burr of diameter 1.5–3.5 mm that is advanced over a thin guide wire. The atherectomy catheter is introduced through a guiding catheter and positioned near the stenosis. The abrasive tip is then rotated at about 150,000 rpm while it is being advanced through the atherosclerotic blockage.

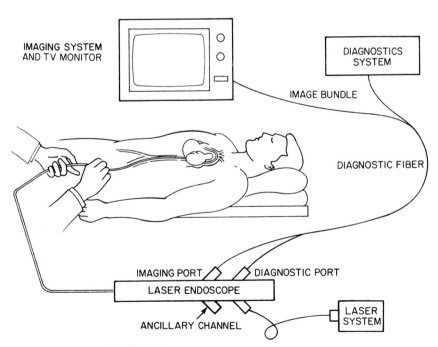

FIGURE 9.2   Laser endoscope system in cardiology.

(iii) *Transluminal extraction* devices (Stack *et al.*, 1991) are based on the excision of small segments of the atheroma and extraction of this atheromatous debris through the atherectomy catheter.

These methods were used successfully to treat blocked peripheral arteries. More than 2000 patients have had coronary atherectomy using mechanical devices, with a high rate of success. Yet all are limited by the catheter diameter, which is larger than that of a laser catheter. Although these devices have been introduced through guiding catheters, in the future it may be possible to insert them through endoscopes.

9.2.3 Endoscopic Imaging—Principles

The cardiologist needs to obtain a good-quality image of the lumen of a blood vessel in order to see the exact shape (and color) of a plaque blockage, which is often asymmetric. Cardiovascular imaging may serve for morphological and pathological diagnosis and will help in various percutaneous interventions. Imaging will enable the physician to direct a laser beam to the exact location of the plaque to be removed, in order to open up a channel in the blockage, while ensuring that the normal blood vessel wall is not affected by the laser beam. Imaging is rather complicated in a thin blood vessel because of the presence of blood that blocks the view and the need to use endoscopes whose diameters are less than 3–4 mm.

9.2.3.1 Angioscopy

The properties of thin and ultrathin fiberoptic endoscopes were discussed in Section 6.3.2. In cardiology there has been interest in carrying out laser angioplasty under angioscopic imaging (Abela *et al.*, 1986). In this section, the particular problems of endoscopic fiberoptic imaging in cardiology are presented.

The system used for guidance and angulation in a thick endoscope is based on mechanical means such as metal wires. Similar methods may be used for thin endoscopes, as shown in Fig. 9.3. In this case the balloon serves to position the thin endoscope at the center of a blood vessel during imaging. It is awkward to

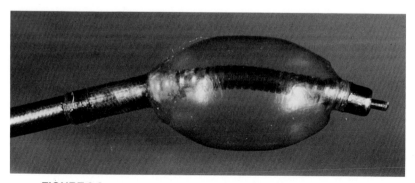

FIGURE 9.3   The tip of a steerable thin angioscope. (Courtesy of Mitsubishi.)

use the same measures for thin endoscopes, and ultrathin endoscopes are currently not steerable. Guidance of the endoscope tip to its exact location was discussed earlier.

The presence of blood makes it difficult to obtain a good image inside a blood vessel. A few methods have been tried to alleviate this problem, such as injecting transparent fluid (e.g., saline or artificial blood) through the endoscope to facilitate viewing. Another method is viewing through a transparent balloon attached to the tip of the endoscope.

During the past few years, there has been progress in angioscopy, especially in the optical quality of the image (Mizuno *et al.*, 1989). Clinical investigations have been performed on the pulmonary artery, cardiac chambers and valves, abdominal and peripheral arteries, coronary arteries, and congenital malformations of the cardiovascular system. Few complications were reported. Coronary angioscopy was used to show features such as narrowing, ruptured atheroma, occluding thrombus, and the lumen surface. Coronary angioscopy was also performed before and after procedures such as percutaneous coronary angioscopy (PTCA) or laser angioplasty (Uchida *et al.*, 1992). The dilatation of the stenotic segments of the arteries was clearly observed, as shown schematically in Fig. 9.4. It depicts a picture taken with the Olympus ultrathin endoscope placed inside the coronary artery, before and after laser angioplasty.

9.2.3.2 Ultrasound Imaging

The most commonly used angiography methods provide two-dimensional images of the lumen. On the other hand, angioscopy provides information about the interior surfaces of blood vessels. Neither method, however, provides information on the thickness of the arterial wall or the thickness and composition of

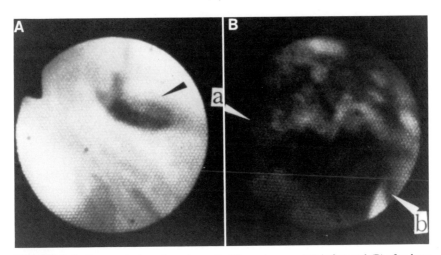

FIGURE 9.4 Images obtained through an ultrathin angioscope: (A) before and (B) after laser angioplasty. (Courtesy of Dr. Y. Uchida.)

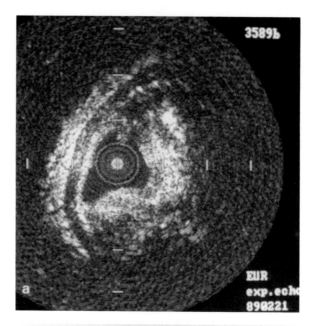

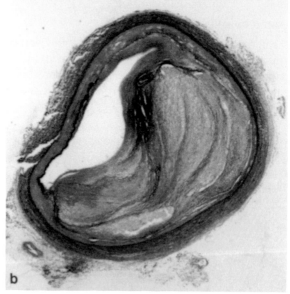

FIGURE 9.5 (a) Ultrasound image and (b) histologic cross section of a femoral artery, following laser angioplasty. (Courtesy of Dr. E. Gussenhoven.)

the atherosclerotic plaque, which is important for laser angioplasty. This knowledge may be provided by miniature ultrasound imaging devices which are attached to the tip of a catheter or an endoscope (Tobis *et al.*, 1989). Such devices have already been tested *in vitro* and *in vivo*. Ultrasound images were obtained from diseased arteries and the luminal cross section, wall thickness, and plaque structure were measured. The results were compared to those obtained by histology. Good correspondence was found between the two methods, indicating that the information obtained by ultrasound imaging may be used in the future for monitoring and control during a laser angioplasty procedure (Bom and Roelandt, 1989).

Figure 9.5 shows the ultrasound (also called echographic) image and the corresponding histologic cross sections of a superficial femoral artery following laser angioplasty. Figure 9.6 shows ultrasound images obtained *in vivo* from a patient

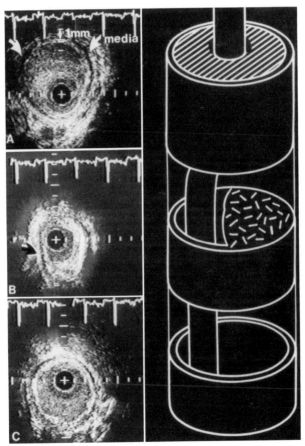

FIGURE 9.6 Ultrasound images obtained *in vivo* from a femoral artery, following laser angioplasty. (Courtesy of Dr. E. Gussenhoven.)

with obstructive disease of the superficial femoral artery after laser angioplasty. The diagram helps illustrate the ultrasonic cross sections.

9.2.4 Diagnosis—Principles

9.2.4.1 Fiberoptic Diagnostics

Chapter 7 discussed a few of the fiberoptic methods used for diagnosis. It was mentioned that a laser beam sent through an optical fiber interacts with blood or with tissue. A returned optical signal sent through the same fiber serves for diagnosis. The optical fiber can be inserted into the body either through a catheter or through a thin endoscope; these methods are well suited for cardiology. They can be used, as mentioned earlier, to monitor blood pressure, blood flow, pH, glucose content in the blood, and so forth. This can be performed either during a regular checkup of a patient or during a laser (or nonlaser) surgical procedure. Several systems have been developed especially for cardiology and are being tested in preclinical and clinical studies.

9.2.4.2 Guidance by "Smart" Systems

Laser-induced fluorescence spectroscopy may be used to distinguish between plaque and normal tissue. Such measurements may be taken through a laser catheter and used to determine the presence (and type) of atherosclerotic plaque. This is the basis of monitoring and control in "smart" laser angioplasty systems.

Scientists (Douek et al., 1991; Garrand et al., 1991) have proposed sending a low-power laser beam (e.g., HeCd UV laser) through a fiber to induce endogenous fluorescence. This fluorescence is collected through the same fiber or through a bundle of fibers. Spectral analysis of the emission functions to identify the tissue in front of the fiber. If this tissue is identified as atherosclerotic plaque, a pulse of a high-power laser beam (e.g., Ho:YAG or dye laser) is sent through the same fiber, or through a special power fiber, to ablate the plaque. The same procedure continues until the tissue in front of the fiber is identified as normal tissue; the ablation process is then terminated. Others (Papazoglou et al., 1990) used UV lasers (e.g., XeCl excimer laser) for ablation and measured the tissue fluorescence through the same fiber. It was found that the fluorescence differs for atherosclerotic plaque and for normal tissue. The laser-induced fluorescence may therefore, once again, serve as a feedback signal which controls the ablation process. The proposed smart system can, in principle, ablate atherosclerotic plaque efficiently without damaging the arterial wall.

A few systems have been tried (Abela, 1990) in vitro and in vivo and several difficulties were encountered. It was found that repetitive high-power laser pulses induce changes in the tissue surface which give rise to changes in the induced fluorescence. Difficulties have also been found in identifying plaque in vivo, because of the presence of blood. In clinical studies, these difficulties led to various complications. Most important, the smart system could not distinguish well between plaque and normal tissue and did not prevent perforations. All these problems will have to be solved before the smart systems become practical.

9.2.5 Fiberoptic Laser Therapy: Angioplasty—Principles

Arteries blocked by atherosclerotic plaque are considered, where we may distinguish between peripheral arteries (in the legs) and coronary arteries (in the heart). A laser beam delivered through an optical fiber serves to remove the blockage. Atherosclerotic plaque is a mixture of fibrous tissue, fat, and calcium that varies not only between patients but within the same patient. It can be soft and easily melted by laser energy. It can also be hard, making it difficult to remove. There are three major concepts for the recanalization of the blocked artery to facilitate blood flow: (i) molding of the plaque, (ii) removal of plaque, and (iii) photochemotherapy. In each case, one of the many lasers and fibers which were mentioned in Table 8.4 is used to perform the laser procedure inside the artery.

This section discusses briefly the guidelines for the uses of different laser–fiber methods. Full details of the laser angioplasty techniques, limitations, and complications are given in the books and review articles mentioned in Section 9.1.

9.2.5.1 Plaque Molding

This method is most applicable in cases in which the plaque is not too hard. The following methods have been tried for this purpose:

(i) Hot tip: This tip was mentioned in Section 8.6. A small metal ball is attached to the tip of an optical fiber and inserted into an artery via a thin catheter. X-ray fluoroscopy is used to guide the metal tip to the plaque. An argon or Nd: YAG laser beam sent through the fiber heats up the metal ball to a temperature of about 600°C in a few seconds. The tip is rapidly pushed through the plaque, which melts and leaves a channel. If the cardiologist keeps the metal tip at one place for too long, it will stick to the blood vessel or even perforate the arterial wall. Although the system has been used clinically, perforations and other complications have been reported (Sanborn, 1988).

(ii) Miscellaneous tips: Various types of tips have been tried in clinical and animal experiments, including attached tips such as flat windows, contact probes, and optical shields (see Fig. 8.3). In other cases, the ordinary tip of the fiber was shaped in the form of a lens, ball, or cone (see Fig. 8.1) (Borst, 1987). All these tips have been tried clinically.

(iii) Laser balloon method: This is a combination of two methods (Spears, 1986). A balloon catheter is used, much as in a regular PTCA procedure, to push plaque to the sides and generate a new channel. Inside the catheter, an optical fiber delivers laser energy. Unlike regular optical fibers, which send the beam in the forward direction, this fiber scatters the laser beam evenly and sideways. The beam must be sent through the balloon while the balloon is inflated. The plaque, which has been expanded by the balloon, is heated and the cracks formed in the plaque are welded. Clinical coronary angioplasty trials are under way (Reis *et al.*, 1991).

9.2.5.2 Plaque Removal

The laser beam is used to vaporize the plaque. The beam is delivered through a power fiber that is inserted either through a catheter or through an endoscope. Several alternative methods can be used. This discussion is divided into three sections; the laser, the fiber delivery unit, and the fiber tip.

Laser

In order to remove the plaque, the laser beam must be absorbed by the plaque or by a coloring agent that has been selectively retained in the plaque. A laser beam that is absorbed in the coloring agent heats up the plaque, causing it to vaporize.

The first laser used for laser angioplasty was the Ar laser. The blue–green light of the Ar laser is highly absorbed by blood and tissue which contains blood. The radiation of this laser, however, is not absorbed well in white or yellowish plaque tissue. The Ar laser beam therefore does not cut plaque faster than it cuts the arterial wall. In addition, this is a continuous-wave (CW) laser that causes thermal damage. Another laser that has been tried is the excimer laser, which has two advantages: the UV is highly absorbed in plaque, and the laser beam is pulsed. As explained in Section 3.8, the pulsed mode leads to tissue removal with little thermal damage. Other pulsed lasers that are highly absorbed in plaque, such as Er:YAG lasers or CO_2 lasers, may also cut tissue with little thermal damage. When the plaque has been colored by dye, a laser with an emission wavelength tuned to the absorption peak of the dye must be used. Tunable dye lasers are the most suited for this purpose.

Fibers

Section 4.8 discussed the various power fibers. With angioplasty, fused silica fibers can be used for near infrared, visible, and the longer-wavelength excimer lasers ($\lambda > 300$ nm). Infrared-transmitting fibers must be used both for the Er:YAG and the CO_2 laser.

Fiber Tips

In some of the experiments reported to date, the distal tip of the power fiber was bare and well polished, but this is likely to cause mechanical perforations. The sapphire tip mentioned in Section 8.3.2, with its rounded form, appears to be much safer.

Blood

The presence of blood presents a special problem in laser angioplasty. The fiber tip can be positioned a few millimeters from the atherosclerotic plaque. There is a layer of several millimeters of blood between the fiber tip and the blockage. Whether an excimer or CO_2 laser is used, this blood layer absorbs nearly 100% of the laser radiation. Several methods have been proposed to solve this problem. Blood can be pushed away with a saline solution. Although this is helpful with visible laser beams (e.g., dye), it does not work with excimer, Er:YAG, or CO_2 lasers, whose radiation is highly absorbed in water. Alternatively, the blood flow can be stopped for 10–20 sec with pressurized gas (e.g., CO_2 gas), as shown schematically in Fig 9.7. The laser beam passes through the gas, vaporizing the plaque. After the 10- to 20-sec period, the blood flow resumes and the procedure is repeated until the blockage is recanalized. The last method may not be safe enough for practical use. When contact tips are used, the tip actually touches the plaque and the blood is displaced.

9.2.5.3 Photochemotherapy

The idea behind this method is identical to the one used for cancer treatment (see Section 3.7.2). A drug is injected into the body and selectively retained by the atherosclerotic plaque. This drug can be injected into the blood stream or directly injected into the diseased area with a catheter or an ultrathin endoscope. The drug is then triggered using a suitable laser wavelength and destroys the host plaque. Preliminary experiments have been conducted using hematoporphyrin derivative (HPD) (Spears, 1986).

9.2.6 Advances in Clinical Testing of Laser Angioplasty Systems

In trying to assess the efficacy of their systems, various research groups chose different laser–fiber systems for their studies. Ultrathin endoscopes are still not widely used. This section discusses the laser catheter systems which have already been tried clinically.

Balloon angioplasty cannot be performed in totally occluded arteries. In principle, the laser procedure is sufficient to generate a new and large lumen in a totally blocked artery. In practice, laser beams were sent through small-diameter optical fibers and generated a "pilot channel." A balloon catheter was then inserted through the narrow lumen and used to enlarge it. This laser-assisted balloon angioplasty procedure, however, suffers from the same high restenosis rate as regular balloon angioplasty.

9.2.6.1 Ar (Argon) Laser Catheter

The Ar ion laser was the first laser tested for angioplasty in animal experiments and preclinical experiments in the early 1980s (Abela *et al.*, 1982; Choy *et al.*, 1984; Lee *et al.*, 1983; Geschwind *et al.*, 1984; Macruz *et al.*, 1980). The lasers were CW Ar lasers emitting visible (blue or green) radiation with power levels of several watts. The fibers used were either regular glass or fused silica fibers. In all the experiments, the distal end of the fiber was bare. Ginsburg *et al.*

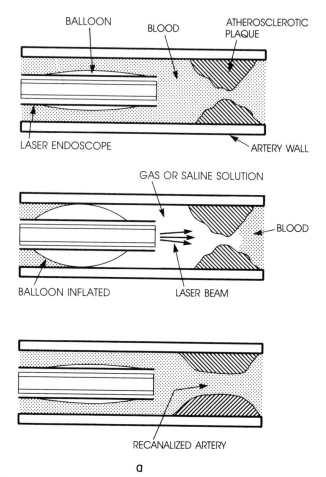

FIGURE 9.7   Laser angioplasty system: (a) schematic drawing and (b) artist's view. (Courtesy of Advanced Interventional Systems.) (*Figure continues.*)

(1985) used the system for clearing blocked peripheral arteries, and Choy *et al.* (1984) experimented with coronary arteries. Severe complications mentioned in these two studies limited their clinical importance.

9.2.6.2 Nd:YAG Laser Catheter

The radiation of the Nd:YAG laser is not highly absorbed in tissue and is therefore not efficient in vaporizing plaque. Yet, a laser catheter based on CW Nd:YAG lasers and "bare" silica fibers was readily available in the early 1980s. This system was thus among the first used clinically for peripheral laser angioplasty. The method was unsatisfactory both because of its inefficiency in removing

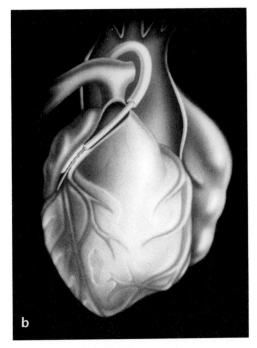

FIGURE 9.7   (*Continued*)

plaque and because of the risk of complications such as perforations. Addition of a sapphire contact tip improved the tissue removal efficiency, and the laser catheter based on CW or pulsed Nd:YAG lasers has been used clinically for laser angioplasty (Linnemeier and Cumberland, 1989).

9.2.6.3 Excimer Laser Catheter

Excimer laser radiation, which is highly absorbed in tissue, can be used to vaporize plaque (including calcified plaque). As mentioned in Section 8.6.3, there were severe problems with the transmission of this laser energy through silica fibers. The early systems, based on thick silica fibers, were too stiff and had to be replaced by a bundle of thinner fibers. A "ring" catheter consisting of 200–300 individual 10-μm fibers in a concentric array with an outer diameter of less than 2 mm (6 Fr) is often used. Many of the technical problems involved in operating XeCl excimer lasers and in using the laser catheter have been solved, making it possible to use this system clinically (Wollenek *et al.*, 1988). Clinical work was limited first to the peripheral blood vessels. More recently, improved catheters have been used either in coronary laser angioplasty or in laser-assisted balloon angioplasty. X-ray fluoroscopic images of a blocked artery before and after laser angioplasty are shown in Fig. 9.8. Thousands of coronary plaques have been re-

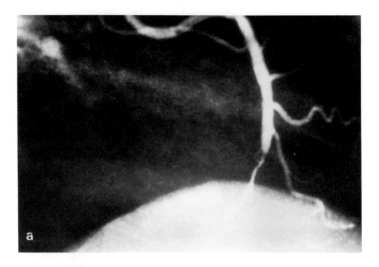

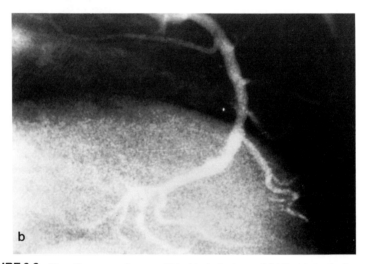

FIGURE 9.8 X-ray fluoroscopy images of (a) a blocked artery and (b) the artery after laser angio-plasty. (Courtesy of Advanced Interventional Systems.)

canalized with this method, with a relatively small percentage of complications. Excimer laser angioplasty was found most useful in stenoses which cannot be crossed or dilated with a balloon, in long segments of diseased arteries, and in calcified lesions (Cook *et al.*, 1991).

Some of the procedures depicted in this section may be described with the help of a flowchart, as mentioned in Section 9.12.

9.3 FIBEROPTIC LASER SYSTEMS IN GASTROENTEROLOGY

9.3.1 Introduction

In the mid-1960s, it was realized that laser energy, when absorbed in blood, causes coagulation. In 1970, Goodale used, for the first time, a laser endoscope to control gastric hemorrhage. Goodale generated bleeding ulcers in the stomachs of dogs, introduced rigid endoscopes into their stomachs, transmitted a CO_2 laser beam through the endoscopes (without a waveguide), and stopped the bleeding. In 1971, the same laser was used in an open operation to stop bleeding in the stomach. In 1973, Nath *et al.* transmitted several watts of Ar laser power through thin (150 μm) fused silica fibers which were inserted in the ancillary channel of a fiberoptic gastroscope. They used this laser endoscope to coagulate intestinal bleeding in a dog. Several groups in Germany and the United States followed these experiments with Ar laser endoscopy (Fruhmorgen *et al.*, 1976), and in 1974, accomplished coagulation of actively bleeding lesions in patients. In 1976 Kiefhaber performed the first experiments using Nd:YAG laser gastroscopes for treating gastrointestinal tract bleeding. He found that the hemostatic properties of this laser were excellent, because of the large penetration depth into tissue (Kiefhaber *et al.*, 1977). Other researchers, notably Brunetaud in France, Bown in England and Dwyer and Fleischer in the United States, used the Nd:YAG laser for similar purposes (Dwyer, 1986).

Laser endoscopy in gastroenterology emerged as one of the most important uses of lasers in medicine (Fleischer, 1987; Hunter, 1989, 1991). It has been used to control bleeding, for the palliation of obstructing esophageal and rectal carcinomas, and for the treatment of gastrointestinal tumors. The following sections discuss some of the diagnostic and therapeutic applications of laser endoscopes.

9.3.2 Gastroscopic Imaging and Diagnosis

Various fiberoptic endoscopes are used in gastroenterology, such as the gastroscope and colonoscope. These endoscopes are among the best developed and have high resolution and excellent color rendition. Special endoscopic techniques (see also Secion 6.5) have also been developed (Sivak, 1987) to facilitate diagnostic imaging:

- Magnifying endoscopes with magnification of $30-150\times$ have been developed to observe the gastric mucosa or colonic mucosa *in situ*. They could, in the future, replace histologic examinations.
- Chromoscopy is based on coloring agents which are applied to tissue through the endoscope and accentuate diseased areas.
- Endoscopic ultrasonography is based on miniature ultrasound transducers which are attached to the endoscope and used for ultrasonic imaging. The images obtained may be used to diagnose the mucosa, the submucosa and the layers of the gastric wall, hepatopancreatobiliary diseases, early evi-

dence of gastrointestinal tract tumors, and the depth of cancer invasion into healthy tissue.

9.3.3 Endoscopic Laser Photocoagulation

Gastrointestinal bleeding is a common cause of emergency hospitalization, with a mortality rate of about 10%. This high mortality rate may be, in part, due to the complications of emergency surgery and may be reduced by less invasive methods. Laser endoscopic photocoagulation offers a less invasive method which is controllable and potentially useful for the management of acute hemorrhage and reduction of recurrent hemorrhage. Laser photocoagulation has been used to treat bleeding from peptic ulcers, esophagogastric varices, benign mucosal lesions, and gastric polyps (Fleischer, 1987).

To treat such bleeding, a laser endoscope is introduced and brought to the vicinity of the bleeding site. One of the most severe problems involves the ability to identify the exact site and nature of the bleeding. Blood must therefore be removed around the distal tip of the endoscope. This is often accomplished by sending pressurized CO_2 gas through the ancillary channel in which the power fiber is located. The excess gas in vented through a second ancillary channel, to prevent discomfort to the patient and gastric overdistention. The gas flow also serves to cool the tip of the power fiber during power transmission and clears secretions from the tip.

Both Ar and Nd:YAG lasers have been used for coagulative laser therapy. The Ar laser emission at 488 and 514 nm is highly absorbed in blood. Its advocates claim that it immediately generates a layer of coagulum that will stop the bleeding. To do this, one needs power levels of less than 10 W and a total of less than 100 exposures, each of duration 1 sec. The lower power is believed to reduce the risk of perforation. The Nd:YAG emission at 1064 nm is not so highly absorbed in hemoglobin. It is transmitted through blood, penetrating deeper into the tissue around the bleeding site, and its advocates claim that it results in better coagulation. One needs power levels of up to 100 W and a total of about 50 exposures of 1 sec duration to stop bleeding. The higher power may also be advantageous when there is a need to control bleeding from large vessels. Sapphire tips have also been used for "contact" application of the laser to the bleeding site.

Many groups have conducted series of controlled and uncontrolled trials with different laser endoscopes under a wide range of operating conditions. There is no question that laser endoscopy can control active hemorrhage, with a high rate of success. It has not been shown conclusively, however, that recurrence of hemorrhage is decreased or that the mortality rate is reduced. More work is needed in this area.

9.3.4 Endoscopic Nd:YAG Laser Therapy

Nd:YAG laser endoscopes have been used to treat malignant tumors (Fleischer, 1987) such as esophageal carcinoma which blocks the esophagus and has a poor prognosis. This disease is accompanied with great suffering to the patient due to dysphagia (inability to swallow) and even palliative treatment would be valuable (Fleischer, 1987). A laser endoscope may be inserted into the esophagus and advanced to the outer margin of the tumor, as shown in Fig. 9.9. Laser energy of several kilojoules (100 W peak power) is transmitted through the power fiber, causing vaporization and thermal damage to the central area of the tumor. After 2 days, the treatment is resumed. The laser-treated tissue is necrotic and is removed through the endoscope. Laser energy is reapplied to a deeper region of

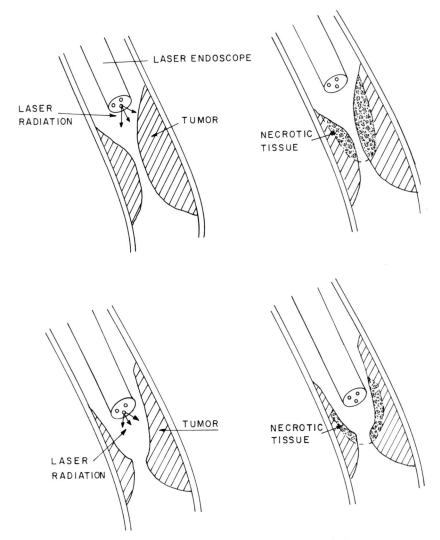

FIGURE 9.9 Laser endoscopy for treatment of gastrointestinal tumor.

the tissue. It again necroses in 2 days and is removed. After several treatments, one may open a new lumen through the tumor, improving swallowing and the quality of life for the patient. It is then possible to insert a stent to hold the lumen patent (i.e., open). Laser recanalization of obstructing tumors is a safe and effective method and is particularly important in cases unsuitable for surgery or radiotherapy.

This method has also been applied to inoperable rectal tumors. The Nd:YAG laser energy opens a channel while coagulating blood vessels. This offers a treatment for the two main symptoms associated with rectal tumors—obstruction and bleeding. The same method has also been applied for colonic tumors.

9.3.5 Diagnosis and Photodynamic Therapy

Photodynamic therapy (e.g., HPD–PDT) is well suited for the treatment of gastrointestinal cancer (Marcus, 1992). In the case of superficial esophageal cancer, early diagnosis and HPD–PDT have been used clinically. HPD–PDT has also been used for palliation of dysphagia that is caused by a malignant tumor. In both cases a cylindrical diffuser tip may be inserted into the esophagus and used for intraluminal treatment. In the case of colon cancer, endoscopic laser surgery or therapy may cause perforations. HPD–PDT is potentially safer. In early clinical studies fiberoptic tips were inserted in colorectal cancer tumors and interstitial treatment was used. In these studies the HPD–PDT was found promising. In principle, the same method will also be useful in the future for early detection and treatment of gastric cancer. But the problems of uniform illumination and of dosimetry are more difficult because of the shape of the stomach, the effects of gastric folds, and the peristaltic motion.

9.4 FIBEROPTIC LASER SYSTEMS IN GENERAL AND THORACIC SURGERY

9.4.1 Introduction

Endoscopic laser surgery, possible to perform for 30 years now, has not been used by many general surgeons (Joffe, 1989). This changed when surgeons started using laser laparoscopic techniques, which had actually been developed for gynecology. The first successful case was laser cholecystectomy, described in the next section. Preliminary work on laparoscopic bowel resection, adhesiolysis, and welding is promising. It is expected that laser surgery of the appendix (i.e., appendectomy) or the liver will follow. As am example, the use of a laser catheter for thoracic surgery is shown schematically in Fig. 9.10.

9.4.2 Laparoscopic Laser Cholecystectomy

The gallbladder, situated on the underside of the liver, contains bile. Stones form in the gallbladder when the bile is oversaturated with cholesterol or bilirubin.

FIGURE 9.10 Endoscopic laser procedure in thoracic surgery. (Courtesy of LaserSonics.)

These stones may block the bile duct that leads from the gallbladder to the small intestine, causing cholecystitis—inflammation of the gallbladder. These stones may be shattered by shock wave lithotripsy or even by laser lithotripsy (Lux *et al.*, 1986). Unfortunately, the stones re-form within a few years and must be removed again. The treatment of choice is often surgical removal of the gallbladder. This cholecystectomy procedure is a major operation which involves a large abdominal scar, a stay of a few days in the hospital, and an extensive recovery period at home.

An alternative solution involves the use of a less invasive laparoscopic cholecystectomy. Typically, four small incisions are needed for this procedure. Through one incision, a rigid laparoscope is inserted, through which the physician obtains a clear image of the gallbladder. Surgical instruments are introduced through two trocar sheaths that are inserted into incisions to hold the gallbladder and other tissues in place. A dissection tool is threaded through the fourth incision and is used to dissect the gallbladder from the liver bed. The first clinical studies used either electrocautery (i.e., coagulating scissors) (Dubois *et al.*, 1990) or a laser catheter (Reddick and Olsen, 1989). The gallbladder can then be removed from the abdominal cavity through one of the trocar sheaths. Electrocautery is a well-established technique that facilitates hemostasis. Yet there have been complications, especially those related to intestinal burns or perforations due to current leakage. The laser beam is easier to control, but the procedure may take longer. It has not yet been established which of the dissection techniques is better (Smith *et al.*, 1990).

Laser cholecystectomy studies have been made with the Nd : YAG laser catheter, using a fused silica fiber and an attached sapphire contact tip. Alternatively, the distal tip of the fused silica fiber itself is specially treated, so it could also be used in contact with tissue during the cutting procedure.

Preliminary studies of laparoscopic cholecystectomy were successful and have generated wide interest (*American Journal of Surgery*, vol. 161, March 1991). This method has been applied successfully on more than 20,000 patients, within 2 years of its introduction. It leads to shorter hospitalization, more rapid recovery, and much better cosmetic results.

9.5 FIBEROPTIC LASER SYSTEMS IN GYNECOLOGY

9.5.1 Introduction

Lasers were introduced into gynecology in the mid-1970s and the first laser operations were carried out by Bellina and French using a CO_2 laser (Bellina 1974; Bellina *et al.*, 1985; Baggish, 1985). They were found to be useful in treating diseases of the lower genital tract such as vulvar, cervical, and vaginal lesions. Internal imaging was first carried out with rigid endoscopes, such as colposcopes, laparoscopes, or hysteroscopes. Laser beams were focused directly on tissue, through rigid endoscopes, and used for endoscopic surgery. Ar and Nd:YAG laser beams were then delivered through power fibers in laparoscopes. Fiberoptic endoscopes have also been widely used in gynecology. One of the most important uses of the CO_2 laser endoscope is in the treatment of cervical intraepithelial neoplasia (CIN), a premalignant lesion of the uterus cervix. Laser endoscopes have also been employed for the management of endometriosis and for intra-abdominal and intrauterine surgery. Because of the early success of laser and fiberoptic systems, gynecology is the medical specialty in which the greatest number of laser procedures are performed in the United States. Some of these endoscopic applications of lasers in gynecology are discussed below. Photodynamic therapy (e.g., HPD–PDT) has also been used clinically in gynecology. Developments related to lasers in gynecology are discussed in several books (Bellina and Bandieramonte, 1986; Keye 1990; McLaughlin, 1987, 1991).

9.5.2 Lower Genital Tract

9.5.2.1 Cervical Intraepithelial Neoplasia

This is a pathological process that results in the formation of neoplasm in the cervix and, in severe cases, may develop to carcinoma *in situ* (CIS). The disease is diagnosed by a cervical smear test that is now routinely used in conjunction with two endoscopic methods: magnification and coloring. The special vaginal endoscopes, called colposcopes, have magnification of $5-25\times$ and are used for examining the cervical epithelium. After the application of an acetic acid solution, irregular epithelial areas become accentuated, making it easier to distinguish the CIN lesion, but this examination is not always definitive. When the results of colposcopy are definitive, however, the lesion is treated effectively by local destruction of the lesion to a depth of a few millimeters.

Several techniques are available for destruction of CIN of the cervix, such as local electrical heating (diathermy) or local freezing (cryosurgery). Effective electrical heating must be performed under anesthesia. Thermal damage to the neighboring tissue may cause complications. The cryosurgery procedure is hard to control and also causes postoperative complications. Vaporization of CIN by highly absorbed laser beams, such as the CO_2 beam, is ideally suited for this problem. The laser beam may be coupled into the rigid colposcope via a long focal length lens and focused to a small spot. The tissue is vaporized with minimal blood loss and pain and with a low complication rate. This laser procedure may therefore be performed without general or local anesthesia on an outpatient basis. Laser endoscopic diagnosis and therapy of CIN has been performed on thousands of patients with a success rate greater than 80%.

9.5.2.2 Vaginal and Vulvar Lesions

Vaginal and vulvar interepithelial neoplasia are much less common than CIN. The traditional surgical procedures in both cases are difficult and may result in complications. Both cases may preferably be treated by the CO_2 laser vaporization through a magnifying colposcope. General anesthesia is recommended. The same method is used now for perineal and vulvar condylomatous lesions, which are benign and bulky tumors caused by viral infection and are contagious.

9.5.2.3 Laparoscopic Laser Surgery

The laparoscope, introduced surgically into the abdomen, is used for examining the pelvic organs. In gynecology, rigid and flexible endoscopes have been used for intra-abdominal laser surgery, as illustrated in Fig. 9.11. A trocar with an outer sleeve (cannula) is introduced into the abdomen; the inner part of the trocar is removed and the endoscope is inserted through the cannula. This endoscope may be a laser endoscope that serves for laser power transmission (via an optical fiber) as well as for irrigation or suction. There are gynecologists who prefer to insert an imaging laparoscope through one puncture, the power fiber through a second puncture and the irrigation/suction tube through a third puncture. In the late 1970s and 1980s, mostly rigid laparoscopes were utilized, but more recently (Cook and Rock, 1991) fiberoptic flexible ones have also been used.

- Endometriosis: Ectopic nests of endometrium may be found in different locations in the abdominal cavity. This often results in the formation of cysts which contain blood and cause chronic, recurrent abdominal pain. These were treated first by an Ar laser beam which was transmitted through a silica fiber (Keye and Dixon, 1983). CO_2 is much more suitable for this purpose because it is highly absorbed by water and is not selectively absorbed by colored tissue. It therefore vaporizes only a thin superficial layer. The CO_2 laser may be operated in the pulsed mode so that the damage to neighboring tissue is minimal. Baggish performed two clinical stud-

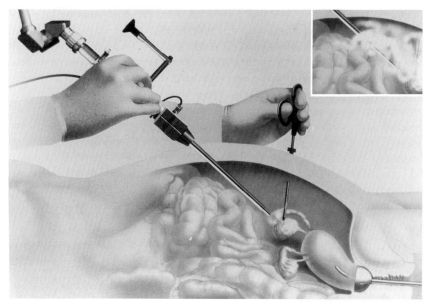

FIGURE 9.11 Laser catheter procedure in gynecology. (Courtesy of LaserSonics.)

ies of laparoscopic treatment of endometriosis, using both rigid and flexible hollow waveguides (Baggish, 1988). The results of the two studies are encouraging.

- Microsurgery for infertility: Laser microsurgery of the fallopian tubes may be performed if disfunction is diagnosed in laparoscopy. Again, this procedure has the advantages of precise tissue removal, bloodless operation, and minimal thermal damage to neighboring tissue. The microsurgical laser operations which have been performed clinically include removal of adhesions (adhesiolysis) and the reanastomosis of the fallopian tubes. The preliminary results of these procedures are also encouraging.
- Intrauterine surgery: The hysteroscope is a special endoscope which has been developed for procedures involving the uterus. It may be introduced into the uterus much like a laparoscope. Laser surgical operations, such as vaporization of the endometrium, have been performed with this endoscope using an Nd : YAG laser and fused silica fibers. An intriguing possibility is the performance of fetal surgery using the same system (Hallock and Rice, 1989).

9.5.3 Diagnosis and Photodynamic Therapy

It is interesting to note that HPD was already used in 1964 for the endoscopic diagnosis of cancer in the cervix (Lipson *et al.*, 1964). PDT is ideally suited for early diagnosis and for the treatment of cancer of the vagina or of the cervix, because they are easily accessible via catheters or endoscopes. Clinical studies have demonstrated that various types of gynecological malignancies can be efficiently eradicated using photodynamic therapy (Marcus, 1992).

9.6 FIBEROPTIC LASER SYSTEMS IN NEUROSURGERY

9.6.1 Introduction

Chapter 2 discussed the general advantages of lasers for surgery and therapy. With the introduction of the operating microscope and its laser adapters and micromanipulator in the late 1970s, lasers made inroads into neurosurgery. For neurosurgery, lasers have the following advantages: (i) the laser beam may be focused to a small area, which is viewed under the magnification of an operating microscope; (ii) the focal spot is easily moved with a mirror system; (iii) the laser beam vaporizes or coagulates tissue in the target area (e.g., a nodule) without mechanical contact and damage to adjacent areas (e.g., neighboring nerves).

Most of the applications of lasers during the past two decades have been intraoperative. CO_2 laser radiation has been used to vaporize tumors in sensitive locations in the brain, pituitary tumors, and meningiomas. Once the exact location of such tumors is determined by computed tomographic (CT) scan, even deeply rooted tumors can be treated in this manner. Spinal cord tumors have also been removed by CO_2 laser radiation, without causing trauma to the spinal cord. Nd : YAG lasers have been used to treat highly vascular and intraventricular tumors, in addition to microvascular repair and vascular anastomosis. These applications paved the way for the use of laser–fiber systems in neurosurgery (Cerullo, 1984; Jain, 1983, 1984; Robertson and Clark, 1988).

9.6.2 Endoscopic Techniques

A few of the endoscopic techniques used in neurosurgery are discussed next.

9.6.2.1 Fiberoptic and Ultrasound Imaging

In the 1920s, endoscopic imaging was already being used in neurosurgery; rigid endoscopes (neuroendoscopes) were used to observe the ventricular cavities or deep-seated tumors in the brain. These have not made a significant impact on neurosurgery.

With the development of CT scanning and magnetic resonance imaging (MRI), it is now possible to determine the location of tumors with great accuracy. A burr hole of diameter 0.5–1 cm is drilled in the skull through which a thin and rigid tube is inserted. A rigid endoscope is inserted through the tube, so that its distal end is brought to the vicinity of the tumor. The endoscope includes irrigation and suction channels and an ancillary channel for auxiliary instruments (e.g., resection or biopsy) or a laser power fiber.

An ultrasound imaging system may also be attached to the tip of the endoscope to allow better positioning of the endoscope. Alternatively, the burr hole is enlarged and a suitable holder attached. A special ultrasound imaging probe is inserted through the burr hole and is positioned in the required position. This probe is then replaced by a rigid plastic tube which, in turn, serves to guide the distal tip of the endoscope to the same position (Otsuki *et al.*, 1992).

9.6.2.2 Endoscopic Surgery

A power fiber is inserted through the ancillary channel of the endoscope. Laser power is transmitted through the fiber and used to treat tumors or other structures. Nd:YAG laser energy transmitted through fibers has been used (Auer *et al.*, 1986) to treat ventricular tumors and cystic tumors in the brain. For cystic tumors, laser energy is utilized to open the cyst and its content are drained through the endoscope. For solid tumors, the procedure involves denaturation of the tissue proteins due to the Nd:YAG laser heating, leading to coagulation necrosis (destruction of tumor tissue).

9.6.2.3 Photoradiation Therapy

Photoradiation therapy is potentially a powerful technique for brain tumors such as gliomas. Because of its selectivity, this treatment may be used to remove tumors in the brain and the central nervous system with minimal disturbance. HPD–PDT was tried clinically on brain tumors in the 1980s. It has been established that HPD accumulates in brain tumors and can be activated with a laser catheter. The power fiber was sometimes introduced into the center of the tumor through a small hole in the skull. In other cases, ordinary surgical methods were used to remove the bulk of the tumor. HPD–PDT was used only on the base of the tumor, in order to eradicate it and prevent tumors from recurring (Marcus and Dugan, 1992). Further studies are in progress.

9.6.3 Lumbar Diskectomy

A common form of lower back pain is sciatic pain, which is often caused by a spinal disk pressing against a nerve root. This condition has been traditionally treated by open surgery. Recently, laser-assisted percutaneous lumbar diskectomy has been used instead. This method is discussed fully in Section 9.9.3.

9.7 FIBEROPTIC LASER SYSTEMS IN ONCOLOGY

9.7.1 Introduction

In previous sections, three distinct ways in which lasers have been used for cancer therapy were mentioned: (i) laser vaporization of malignant tumors, (ii) laser heating for hyperthermia or for coagulation necrosis, and (iii) photoradiation therapy. All three modalities are termed cancer phototherapy and are discussed in review papers (Bown, 1983; Gomer *et al.*, 1989; Dougherty *et al.*, 1990;

Marcus, 1992) and books (Doiron and Gomer 1984; Dougherty, 1989, 1990, 1992; Henderson and Dougherty, 1992; Kessel, 1990; Morstyn and Kayo, 1990).

In this section we discuss briefly the general use of fiber–laser systems for cancer diagnosis and phototherapy. In addition, each section that addresses a specific medical discipline also includes a section on HPD–PDT, the photodynamic therapy based on hematoporphyrin derivative.

9.7.2 Laser Vaporization of Malignant Tumors

Laser energy transmitted through optical fibers in laser catheters or laser endoscopes is able to vaporize or cut malignant tissue. The most suitable lasers for this procedure are the pulsed CO_2, excimer, and Er:YAG lasers, which remove tissue efficiently. Although tumors may also be cut with mechanical tools which are inserted endoscopically, laser vaporization causes much less damage to the surrounding tissues. Laser methods cause less spreading of the malignant cells to healthy regions than mechanical tools. Laser vaporization is a useful technique for the treatment of early malignant disease or for palliative removal of large tumors.

Laser vaporization has been used in gynecology and in laryngology for treatment of early malignant diseases. The main limitation of this method is that the pulsed laser treatment does not coagulate the blood. If bleeding occurs, one must use CW lasers, such as Nd:YAG laser or CO_2 lasers, to stop the bleeding.

9.7.3 Photodynamic Therapy

The use of HPD in photodynamic therapy was studied in the early 1960s (Lipson *et al.*, 1961). HPD–PDT is a viable therapeutic modality for the treatment of cancer (see Section 3.7.2). Malignant tumors respond to this modality even after failure of other modalities such as radiation therapy or chemotherapy. In a worldwide series of clinical trials involving thousands of patients, a variety of tumors were successfully treated. Some of the tumors were in advanced stages of malignancy, and others were in earlier stages. These studies involve many disciplines, which are mentioned in other sections of this chapter. The trials have demonstrated that laser photochemotherapy of tumors may have a fundamental impact on cancer therapy (Dougherty *et al.*, 1990). This section reviews laser–fiber aspects.

- Sensitizer: HPD is given to patients intravenously in a minimum dose of about 3 mg/kg body weight. After 48–72 hr, the sensitizer is localized in malignant tumors. Other sensitizers are also being investigated.
- Diagnosis: HPD may be injected intravenously. After 24–48 hr, the compound is concentrated in malignant tumors. A whole area may now be illuminated by a UV laser beam. The physician may look at the whole area through glass filters that transmit only red light (600–700 nm). A region emitting red light involves a malignant tissue. In dermatology, there is a need to identify basal cell carcinoma. HPD is injected into the body and, after a few days, the face of the patient is illuminated with UV laser light.

The malignant cells are clearly seen by their red emission. In practice, the emission can be recorded using an imaging system with an image intensifier (Andersson-Engels *et al.*, 1990).

- Therapy: When the tumor is exposed to red laser light, the HPD is photoactivated; it converts triplet oxygen to singlet oxygen, which is cytotoxic and kills the tumor cells. The minimum threshold for photoactivation is about 15 mW/cm^2 and a typical fluence is 100 J/cm^2. The exact site of action of the photodynamic therapy is under intensive study. There seem to be effects on both the cellular level and the vascular level. When a malignant tumor has been detected using the laser endoscope, the tip of the power fiber is placed next to or inserted inside the tumor. Red light from an argon pumped dye laser or Au laser is transmitted through the power fiber. This red light photoactivates the sensitizer, which in turn causes a

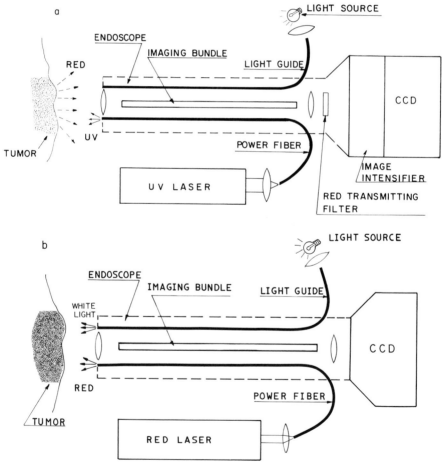

FIGURE 9.12 Endoscopic photodynamic therapy: (a) diagnosis and (b) (HPD–PDT) treatment.

controlled lysis (gradual destruction) of the tumor cells. Figure 9.12 illustrates endoscopic diagnosis and therapy with HPD–PDT.

Figure 9.13 shows endoscopic photographs obtained from a patient with squamous cell carcinoma obstructing the mainstem bronchus.
The general procedures for diagnosis and therapy using HPD–PDT may be described by a flowchart, as discussed in Section 9.12.

9.7.4 Laser Photocoagulation

Laser energy, absorbed in malignant tissue, causes death of tumor cells with delayed sloughing and has been used for cancer treatment. The Nd:YAG laser is

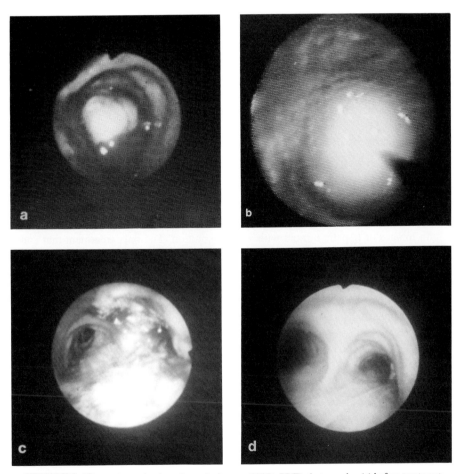

FIGURE 9.13 Endoscopic photodynamic therapy (HPD–PDT) photographs: (a) before treatment (tumor is seen); (b) during laser therapy; (c) "cleanup" after treatment; (d) 3 months after treatment (tumor disappeared). (Courtesy of Dr. S. Lam.)

most suitable for this application because its light is easily transmitted through fused silica fibers and it deeply penetrates tissue. This is particularly important in cases which are unsuitable for conventional surgery or radiotherapy. Nd:YAG laser endoscopes have been used for the palliative recanalization of large malignant tumors which obstruct the upper and lower gastrointestinal tract (Bown *et al.*, 1986) or the main airways (see Section 9.3.4). In regular laser endoscopy procedures, the distal tip of the laser endoscope is brought to the vicinity of the tumor and the distal tip of the power fiber is kept a few millimeters from the tumor. The tip of the power fiber may be inserted inside the tumor and the laser energy is then applied interstitially. This method has also been proved to be effective for the destruction of metastasis in the liver.

9.7.5 Laser Hyperthermia

Sections 3.7.1 and 7.5.1.2 described hyperthermia cancer treatment that is based on heating tissue to temperatures between 42.5 and 43.5°C for tens of minutes. Under these conditions malignant tissue is destroyed while normal tissue is not affected. In laser hyperthermia, the heating is performed by laser energy—normally using an Nd:YAG laser beam which penetrates deep into tissue. The laser energy is delivered through a power fiber which is incorporated in a laser catheter or laser endoscope (Daikuzono *et al.*, 1988). As in laser photocoagulation, the distal tip of the power fiber is placed a few millimeters from the tumor. Alternatively, a diffusing fiber tip is inserted into the tissue for interstitial treatment. Several fibers may be inserted into the tumor with their tips placed (endoscopically) at various depths, facilitating a more even temperature distribution inside the tumor. As an example, cancer of the liver may be considered. Using improved diagnostic methods such as ultrasound or CT, it is now possible to obtain images of hepatic tumors with good resolution. The only treatment that can currently cure such tumors is surgery. Interstitial laser hyperthermia offers an alternative, and by using ultrasound imaging it is even possible to follow the dynamic changes in the tumor during heating and the ensuing necrosis. This method has been tried clinically for cancer of the liver and of the pancreas (Steger, 1991) and it is potentially useful for a variety of other tumors such as those of the adrenal or prostate glands.

9.8 FIBEROPTIC LASER SYSTEMS IN OPHTHALMOLOGY

9.8.1 Introduction

Because the eye is transparent to visible, near-UV, and near-IR light, its inner parts are easily accessible to laser light (see Section 2.4.1). Lasers have thus been used in ophthalmology for therapy and diagnosis for almost three decades. Some of the important laser applications in ophthalmology are briefly discussed.

It has been known for millennia that sunlight can cause damage to the eye. In the 1940s Meyer Schwickerath tried to use concentrated sunlight to produce burn scars inside the eye and found that scars that were produced near holes in the retina prevented retinal detachment (Meyer Schwickerath, 1949). In the 1950s he collaborated with researchers in the Zeiss company in Germany to produce a high-pressure xenon lamp that was used as an intense source of light for retinal "welding." The major problem of this instrument was that this was an ordinary light source. Therefore the light was focused by the eye on a relatively large spot. The irradiance (power density) at that spot was not high and it took a long time (e.g., 1 sec) to produce a burn. During this time heat diffused to other parts of the eye, causing pain and unavoidable eye movement (with a change of position of the spot). Another problem was that the spectral emission of the xenon lamp (at 800–900 nm) was not well matched to the absorption of the retina or the underlying choroid.

With the development of the first ruby laser in the early 1960s, ophthalmologists immediately tried to use it for the same purpose (Ross and Zeidler, 1966). The collimated and monochromatic laser light was focused by the eye to a small spot (<0.1 mm) and lesions were obtained using short pulses (<1 msec), without pain to the patient and without eye movement. The successful use of the ruby laser for intraocular applications paved the way for the use of other lasers for therapy and diagnosis (L'Esperance, 1983).

In ophthalmology therapy, the heating effects of laser beams (e.g., Ar, Kr, or dye lasers) inside the eye may cause photocoagulation. This has been extremely useful for the management of treatment of retinal detachment. A hole or a tear that develops in the retina may lead to separation of the retina from the underlying choroid. The focused laser beam is used to generate around the hole a circle of burn scars that unite the retina to the choroid and prevents detachment. The focused laser beam can also cause heat shrinkage that seals blood vessels inside the eye. This has been used for the treatment of proliferative diabetic retinopathy, a degenerative disease of the retina. Tumors inside the eye are accessible to photochemotherapy treatment, similar to that described in Section 9.7.3, where a suitable red laser (e.g., Au vapor laser) is used. Photodisruption appears when high peak power laser pulses (e.g., Nd : YAG) are focused inside the eye. This has been used for intraocular incision in the posterior capsule of the lens or for incision of vitreous structures inside the eye. More recently, excimer laser beams have been used to reshape the cornea by ablating material from the outer surface of the cornea. Short pulses of highly absorbed laser wavelengths can potentially remove corneal tissue with practically no thermal damage. This refractive keratotomy is potentially an extremely useful method for correcting refractive defects of the eye.

In diagnostics, a laser (e.g., HeNe) beam illuminates the retina and either the scattered light or the electroretinographic response is measured. If the laser beam is scanned across the retina, the physician may detect visual defects or obtain localized functional abnormalities. Visible lasers may also be used for laser Doppler velocimetry (see Section 7.7.1) of the blood in the retinal blood vessels.

These applications of laser do not normally make use of fiberoptic techniques. Some of the special cases in which optical fibers are needed are discussed next (Thompson *et al.*, 1992).

9.8.2 Ophthalmological Applications of Laser—Fiber Systems

In the late 1970s and early 1980s, there were several attempts to perform CO_2 laser surgery or photocautery inside the eye (Bridges *et al.*, 1984; Meyers *et al.*, 1983; Miller *et al.*, 1981). The waveguide used in most of these cases was a rigid metallic or dielectric tube of diameter 1–1.5 mm and length 20–30 mm. A ZnSe window was attached to its distal end. The proximal end of the rigid waveguide was attached to the tip of an articulating arm system (see Section 2.3.4). The distal end was inserted into the eye and placed in the immediate vicinity of a diseased tissue. More recently, other lasers have been used. Some of these applications are discussed:

Photocautery

Miller *et al.* (1981) performed photocautery experiments on rabbits to demonstrate that CO_2 laser energy could be used to close blood vessels and seal retinal tears. They found that, at moderate laser powers, the laser probe performed rather well when the power output was 0.5 W for a duration of 2–5 sec. They then performed several successful clinical trials for closing retinal vessels and sealing retinal tears in the case of severe diabetic retinopathy.

Transection of Vitreal Membranes

Vitreous membranes in the eye may be cut (i.e., vitrectomy) by mechanical tools which are inserted via hypodermic needles. The membranes may then be pulled out through the same ducts. Intraoperative complications may arise if the membranes are too close to the retina, particularly, if the membranes are vascular and bleed into the eye. It may thus be advantageous to insert a probe into the eye and bring it to the vicinity of the membrane. CO_2 laser energy is then delivered and microsurgery is performed for the membrane removal. Laser energy is highly absorbed by the membrane and the surgical procedure may be performed without causing damage to the retina.

Such vitrectomy experiments were performed using CO_2 lasers and rigid (Miller *et al.*, 1981) or flexible (Meyers *et al.*, 1983) probes. More recently, Margolis *et al.* (1989) used a pulsed Er:YAG laser (2.94 μm) and IR-transmitting ZrF fibers of diameter 0.3 mm to cut dense membranes which are adjacent to the retina. The Er:YAG laser radiation, which is also highly absorbed in tissue, offers the same advantages as the CO_2 laser. No retinal lesions were produced if the retina was more than 2 mm from the tip of the fiber. Yet the fiberoptic delivery systems used require further improvements.

All these experiments proved the potential of highly absorbed laser beams for practical use in cutting avascular and vascular vitreous membranes. Further experimentation is needed before they are used clinically.

Cataract Surgery

In cataract formation the lens of the eye becomes opaque, leading to blindness. In cataract surgery the lens is removed through a small incision, leaving the capsule that surrounds the lens. The empty capsule is then filled with a polymer that has optical and mechanical properties similar to those of the lens. Instead of surgery, a laser catheter may be inserted through the cornea into the lens. A laser beam sent through the catheter will be used to vaporize the lens material and the vapors could be pumped out through the catheter. This procedure is likely to be less traumatic than regular surgery. Highly absorbed UV or IR lasers may be most suited for this application and experiments are under way (Thompson *et al.*, 1992).

Glaucoma Surgery

Glaucoma is a disease that is manifested by high pressure inside the eye and may lead to blindness. This pressure may be reduced by generating a small hole in the peripheral iris and facilitating better fluid flow. Traditionally, this hole has been generated by a noninvasive procedure: Ar or Kr laser beams have been focused onto the iris. These beams are highly absorbed by the melanin in the iris, and this leads to ablation of a hole. It may be advantageous in some cases to use a laser catheter for this applications (Thompson *et al.*, 1992).

9.9 FIBEROPTIC LASER SYSTEMS IN ORTHOPEDICS

9.9.1 Introduction

A large variety of tissues are encountered in orthopedics, such as bone, cartilage, ligament, fibrocartilage, muscle, synovia, and tendon. These tissues widely differ in their functions, density, and consistency. There is also a diversity of disorders which orthopedists must address. For centuries, orthopedic surgeons have been using mechanical tools such as saws, drills, chisels, and scissors for cutting and other mechanical devices such as screws, pins, rods, and staples for fixation of tissue.

The efficacy of the mechanical instruments is very high for most surgical procedures. Yet there are cases in which improvements are needed. With all cutting devices, force is applied on tissue (normally by a sharp edge of the instrument), giving rise to tissue separation. There is no real monitoring of the applied force and its control depends on the surgeon's experience. In delicate situations, errors are unavoidable and may lead to complications. In addition, the diseased area is frequently situated inside the body and access to it requires surgical exposure, which is undesirable. In principle, the surgeon may use special mechanical tools which are inserted through a rigid endoscope (arthroscope). In practice, however, it is rather difficult to cut dense tissue with such miniature instruments.

The less invasive methods offered by lasers and optical fibers are likely to change the situation. Highly absorbed laser beams, such as excimer, CO_2, or Er:YAG beams, can ablate tissue with great precision. Tissue ablation is carried out without introducing vibrations or mechanical pressure and with little damage to adjacent tissues. This applies to cartilage, tendon, and even bone. Nd:YAG or CO_2 laser energy may also be used for tissue welding. The laser energy may be transmitted to intracorporeal structures via optical fibers or other waveguides, and visualization may be provided through fiberoptic endoscopes. Some of the developments in this area have been discussed in books and review articles (Sherk, 1990; Whipple, 1987).

9.9.2 Arthroscopic Surgery

In the past, the common methods for treating disorders of the knee or the shoulder involved open surgery of the joint (i.e., arthrotomy), a major surgical procedure. Methods were sought for reaching the confined spaces between the articular surfaces of the joints in the knee or the shoulder with less extensive dissection. Endoscopic (i.e., arthroscopic) techniques make it possible to carry out least invasive interarticular surgery. Mechanical devices are introduced through the arthroscope and used in surgical procedures such as meniscectomy or synovectomy. The diameter of these mechanical devices was normally limited to 3–4 mm, and with such thin instruments it was not easy to perform resection of dense tissue. Manipulation of the mechanical instruments may also cause damage to the contiguous tissue. It was natural to try to replace the mechanical tools with lasers, as illustrated in Fig. 9.14.

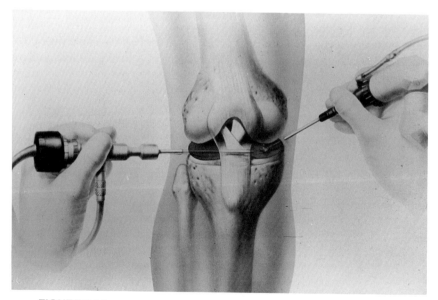

FIGURE 9.14 Arthroscopic laser surgery inside a knee. (Courtesy of LaserSonics.)

The CO_2 laser beam is highly absorbed in human tissue, including meniscus tissue or cartilage. The CO_2 beam, and in particular a pulsed beam, is therefore most suitable for the resection of human meniscus. Preliminary meniscectomy experiments were carried out clinically using a CO_2 laser beam with a special articulated arm. A cannula was attached to the tip of the articulated arm and a tiny mirror was attached to the tip of this cannula. The mirror was used to steer the laser beam. The cannula was inserted into the arthroscope and gaseous nitrogen was introduced through the same cannula in order to obtain distention of the joint space and to flush away the debris. These experiments proved that the meniscus tissue can be removed efficiently without significant thermal damage to neighboring areas, and the remaining peripheral rim of the meniscus heals well. On the other hand, the procedure was limited due to the use of a rigid cannula in a rigid arthroscope. Clinical experiments were also performed (Sherk, 1990) using a CO_2 laser and hollow dielectric waveguides. These procedures offer an attractive alternative to conventional arthroscopy, but they are still somewhat limited by the inflexibility of the waveguide. This situation may improve with the development of flexible IR fibers for the transmission of CO_2 laser beams.

Other lasers such as the excimer, the Nd:YAG, and the Ho:YAG have also been tried clinically for arthroscopic applications. As an example, the Ho:YAG laser energy was delivered through a fused silica fiber; this laser's wavelength is highly absorbed in tissue and cuts the meniscus tissue efficiently. The beam can easily vaporize synovium tissue or loose bodies in the joint. At the same time, the laser beam has hemostatic properties which help prevent postoperative bleeding. In preliminary series of clinical experiments, the fiberoptic-assisted laser arthroscopy was compared to conventional arthroscopic procedures (which make use of mechanical tools). The laser procedure showed significant advantages, in terms of both interoperative effects such as scuffing or trauma and postoperative bleeding or joint inflammation.

9.9.3 Laser Diskectomy

A common orthopedic disorder which affects almost 5% of the population is lower back pain resulting from a herniated intervertebral disk. A ruptured or distorted disk extends into the spinal canal and presses against the spinal cord. This disorder was previously treated by open back surgery such as laminectomy, removal of the lamina, or diskectomy, removal of the disk to reduce the pressure and the pain (cord decompression). Other techniques have been developed in order to reduce the trauma and postoperative effects of intervertebral disk surgery. Some methods involve percutaneous diskectomy using mechanical devices and other methods involve injection of chemicals (e.g., chymopapain) which cause enzymatic removal of the disk. Many of these methods cause complications, including recurrence, and their efficacy is also limited (Sherk, 1990).

During the last decade there have been attempts to use laser–fiber systems for diskectomy. In some early clinical studies (Choy *et al.*, 1987), a fused silica fiber

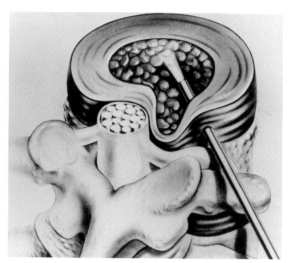

FIGURE 9.15 Laser diskectomy. (Courtesy of Coherent.)

was introduced into the disk space and an Nd:YAG laser beam was transmitted to vaporize tissue and decompress the cord. More recently, a silica fiber was introduced into the nucleus of the disk and Ho:YAG laser energy, which was sent through this fiber, vaporized a portion of the disk tissue (Sherk, 1990). This procedure is illustrated schematically in Fig. 9.15.

The same procedure has also been done using an Ho:YAG laser endoscope system which provided both visualization and guidance during the percutaneous diskectomy procedure. This endoscopic system may afford precise laser ablation and control the volume of nucleus to be removed. In other experiments (Buchelt *et al.*, 1992), an excimer laser beam sent through fused silica fibers ablated diskal tissue *in vitro*. In principle, laser-induced fluorescence (LIF) may be added as a diagnostic tool to distinguish between the two parts of the disk: the annulus fibrosus and the nucleus pulposus. Such a system may be used as another "smart" system that is guided by fluorescence and may prevent penetration of the annulus and injury to the nervous system.

The laser diskectomy procedures are performed percutaneously with local anesthesia. They offer a number of advantages, such as reduced surgical risk, reduced trauma and pain, shorter hospital stay, and faster recovery. When fully developed, laser diskectomy is likely to replace the other procedures described above.

9.9.4 Tissue Welding

While treating the spinal column or the extremities, the orthopedic surgeon is often required to treat disorders of neurological or vascular tissue. When necessary, fixation of these tissues is performed with the assistance of foreign materials such as sutures or glues. Laser welding of tissue has been discussed in detail in Section 3.7.1.3. In particular, laser anastomosis of blood vessels and repair of nerves have been tried using different lasers and various irradiation conditions. Laser anastomosis of blood vessels is potentially useful in orthopedic surgery. It may assist in the transfer of skin or muscle flaps or in the replantation of severed digits or extremities. Nerve repair or grafting is critical to the success of many orthopedic surgical procedures. The laser offers great advantages over conventional suturing techniques. Laser welding is in principle faster, easier to perform, and more reliable. Much more research is needed, however, before this technique is used clinically.

9.10 FIBEROPTIC LASER SYSTEMS IN OTOLARYNGOLOGY (ENT)

9.10.1 Introduction

Natural openings provide easy access into the ear, nose, and throat (ENT). Simple optical systems which have been used for generations made it possible to illuminate and obtain clear images of internal parts. Mechanical tools such as grasping or cutting tools could also be easily introduced into these natural openings, as could electrosurgical and cryosurgical tools. The electrical and cryogenic surgical tools result in uncontrollable thermal damage to neighboring tissue, which is often unacceptable in otolaryngology. For example, in laryngeal surgery, cryosurgery may result in excessive slough and even obstruction of the larynx. Lasers, such as the CO_2 or excimer lasers, are preferable because they remove tissue in a controlled way and with little thermal damage. Laser beams, which are simply reflected by a mirror and focused inside the ear or the oral cavity, have been used for surgical applications.

With the development of rigid endoscopes, it was possible to obtain excellent images of deeper zones in the larynx or the bronchus. Laser beams sent through these endoscopes have been used for surgery and therapy. The introduction of fiberoptic laser endoscopes has changed the situation again. It is hoped that these will make a great impact on endoscopic laser surgery and therapy in otolaryngology and chest medicine. Section 9.10 discusses the topics which are related to fiberoptic laser systems. (Carruth, 1983; Carruth and Simpson, 1988; Davis, 1990)

9.10.2 Endoscopic Laser Surgery—Larynx, Pharynx, and Oral Cavity

In the mid 1970s, lenses were used to focus CO_2 laser beams at a typical distance of 20–30 cm from the lens. Focused laser beams, transmitted in air, were coaxially aligned with rigid laryngoscopes and used for endoscopic surgery (Strong *et al.*, 1973). Many groups used similar laser endoscopes clinically in laryngeal operations such as the treatment of vocal cord nodules (e.g., "singer's nodules"), laryngeal polyps, or other benign tumors such as respiratory papillomas (Shapshay, 1987). The CO_2 laser beam was sometimes used to excise these lesions, which were then extracted with endoscopic forceps. In other instances, the lesions were vaporized until healthy tissue was reached; the rate of success in all these procedures was high. Laser endoscopy was also tried for the management of laryngeal stenoses by endoscopic resection. With benign stenoses, the laser can vaporize or excise the lesions; however, the recurrence rate is not improved in comparison to nonlaser techniques. For malignant tumors, such as carcinoma of the larynx or vocal cord carcinoma, the laser can also serve to excise the tumor. The cure rate is comparable to that of the more traditional surgical procedures or radiotherapy, with less morbidity.

CO_2 laser endoscopy has been also used for the management of benign or malignant tumors in the oral cavity or pharynx and for nasal surgery. This method is useful for the management of both benign and malignant tumors. The benefits of the treatment, in the ease of operation, are the bloodless field and better healing compared to standard surgery.

9.10.3 Endoscopic Laser Surgery—Tracheobronchial Tree

The development of thin, flexible fiberoptic endoscopes has changed bronchoscopy. It is now possible to obtain clear images inside the tracheobronchial tree, perform diagnosis, and use the fiberoptic bronchoscope as a laser endoscope (Shapshay, 1987). Few of the tumors in the trachea and the bronchus are benign, and management of these tumors (as well as stenoses) is similar to that explained in the previous section. The major requirement is early diagnosis of malignant disease and its treatment.

Early diagnosis of carcinoma of the bronchus is carried out by sputum cytology, which shows the presence of malignant cells in the sputum. This method, however, does not provide information regarding the location of the tumors. When the tumors are small, they cannot be easily observed in regular (white light) bronchoscopy and cannot be detected by chest x-ray study. Such early tumors can be detected by the fluorescence techniques described in Section 9.7.3.

Carcinoma of the bronchus may be treated by radiotherapy or by surgery. One of the severe problems is that a significant fraction of the patients suffocate due to blockage of the bronchus. Because of the accumulation of eschar layers, repetitive surgical resection cannot alleviate this problem. Various groups tried to solve the problem with laser endoscopes. The CO_2 laser was used in conjunction with rigid bronchoscopes for laser surgery. Ar or Nd:YAG laser beams were transmitted by fused silica fibers and used in both rigid and flexible bronchoscopes for endoscopic laser coagulation (see Section 9.3.3). Endoscopic laser surgery and therapy are effective as a palliative treatments, improving the life quality of the patients.

9.10.4 Diagnosis and Photodynamic Therapy

The combined use of drugs and laser excitation for diagnosis and therapy, such as HPD–PDT, was discussed in full in Section 3.7.2. This is particularly important in the diagnosis and treatment of carcinoma of the bronchi. As mentioned, early detection of malignant tumors in the bronchi is important and sputum cytology does not provide information about the location of the tumors. Advances in fluorescent diagnosis and endoscopy may provide the necessary answer. Two or three days after HPD injection, a laser catheter is inserted into the bronchial tree and delivers Kr laser light (410 nm). The fluorescence emitted by bronchial tissue is sent back through an optical fiber and passes through a red filter that transmits a narrow band of wavelengths at approximately 630 nm. Red luminescence will be observed only if there is a malignant tumor which was exposed to the Kr laser light. The laser catheter provides a preliminary warning of the presence of tumors, but their exact location can be determined only with fluorescent endoscopy. A laser endoscope may be inserted into the bronchial tree and Kr laser light delivered through the power fiber. Under this illumination, the malignant tumors fluoresce in a characteristic red light which may be observed when a red-transmitting filter is used. The red emission is rather low and an image intensifier is often used to obtain a better picture. Several groups (e.g., Balchun and Doiron, 1982) performed clinical studies and showed that this technique can provide an accurate and quick method for the early diagnosis of carcinoma in the tracheo-bronchial tree.

When malignant tumors have been detected, the same laser endoscope can be used for photodynamic therapy. A red (630 nm) laser beam is sent through the power fiber to illuminate the malignant tumors. The tip of the fiber may be a cylindrical diffuser tip that is placed inside the lumen. It illuminates the inner surface and is used for intraluminal treatment. Alternatively, the tip may be inserted into the tumor to provide interstitial treatment. Endoscopic photographs of HPD–PDT in otolaryngology were shown in Fig. 9.13. Clinical studies (Dougherty *et al.*, 1990; Marcus, 1992) show the enormous potential of photodynamic therapy in the management of endobronchial lung cancer.

9.10.5 Thin and Ultrathin Laser Catheters and Endoscopes for Diagnosis, Surgery, and Therapy

In Chapter 6 we discussed the development of thin and ultrathin fiberoptic endoscopes, some of which have a flexible distal tip. With these one can perform atraumatic endoscopy of the nasal cavity, the paranasal sinuses, the middle ear, and the eustachian tube.

The passages in the upper airways (e.g., the nose) the paranasal sinuses and the nasopharynx are narrow and not easily accessible. The thin laser catheters and laser endoscopes made it possible to carry out diagnosis, therapy, and surgery inside these passages. Figure 9.16 illustrates the use of a laser catheter for laser surgery inside the nose.

Laser pulses sent through suitable fibers can be used to shatter urinary stones, as mentioned in Chapter 3 and discussed in Section 9.11.4. This technique may also be used to shatter stones in the salivary glands (i.e., sialolithiasis). This is another example of a least invasive procedure that will replace a surgical operation

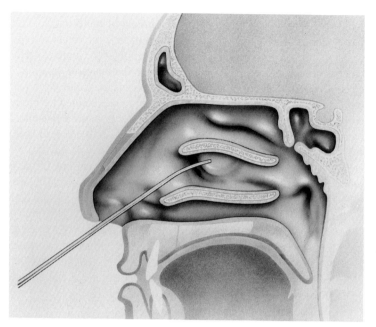

FIGURE 9.16   Endoscopic laser surgery in otolaryngology. (Courtesy of Coherent.)

(Tschepe *et al.*, 1992). Laser lithotripsy in otolaryngology is illustrated schematically in Fig. 9.17.

9.11 FIBEROPTIC LASER SYSTEMS IN UROLOGY

9.11.1 Introduction

For decades, rigid endoscopes have been used in urology and, in particular, for imaging the urethra, the interior of the bladder, and the ureter. Mechanical and electrosurgical systems introduced through these endoscopes have been used successfully for tumor removal, as in the treatment of transitional cell carcinoma of the bladder. Electrohydraulic and ultrasonic devices have also been used for endoscopic therapy of ureteral stones. Laser–fiber integrated systems have shown great promise in the diagnosis and treatment of urological disorders (Staehler *et al.*, 1976).

Lasers have been used for local surgery on the external genitalia, such as the excision of condylomatous lesions of the penis with the aid of the CO_2 laser. In comparison to regular surgery, the proper choice of wavelength and operating conditions gives rise to cosmetically superior results. Lasers have also replaced

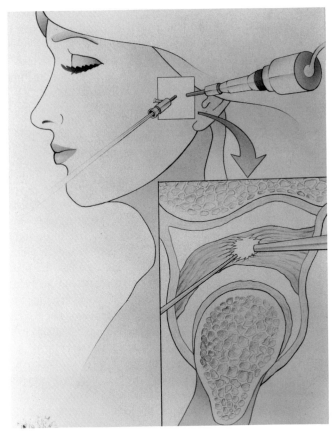

FIGURE 9.17   Endoscopic laser lithotripsy of salivary gland stones. (Courtesy of Coherent.)

surgical knives in open urological surgery. In this section, applications which make use of lasers and optical fibers are considered. Some of the developments in this area are discussed in several books (Hofstetter and Frank 1980; McNicholas 1990; Smith *et al.*, 1990; Steiner, 1988) and review articles (Hofstetter, 1986).

9.11.2 Fiberoptic Laser Surgery in Urology

9.11.2.1 Urethral Stricture

Urethral strictures are usually managed by regular dilatation, by cold knife resection, or by electrocautery resection. The recurrence rate in these methods is rather high because of scarring. Further resection of the tissue may cause further scarring. Stents have also been found useful for the treatment of strictures. Alternatively, a thin laser catheter may be inserted into the urethra; a laser beam delivered through the power fiber in the catheter can, in principle, remove the urethral strictures with no subsequent stricture. Nd:YAG laser radiation was first used for this application, and the deep penetration of the laser energy guaranteed coagulation necrosis throughout the stricture. Later work made use of Ar laser energy, which did not penetrate deeply into the tissue. Although both resulted in a short-term improvement, the strictures recurred. It is possible that pulsed CO_2 laser or excimer laser beams will remove the stricture tissue without damage to adjacent tissue and will result in no scar formation and recurrence.

9.11.2.2 Bladder Tumors

There is a great need for transurethral treatment of carcinoma of the bladder. The goal is to remove completely and destroy the tumor in the bladder wall without perforating the wall and without damage to adjacent structures (e.g., the intestines). Endoscopic resection of the tumors with electrocautery is not satisfactory because of unpredictable thermal necrosis effects and the risk of perforating the bladder wall. These problems can be solved if the tumors are destroyed by Nd:YAG laser energy. The Nd:YAG laser beam was delivered into the bladder via laser catheters, rigid endoscopes (cystoscopes), or fiberoptic endoscopes. Several groups (Hofstetter and Frank 1980; Smith, 1986) have shown that Nd:YAG laser energy, which penetrates through the tumor, causes coagulation necrosis throughout the full thickness of the bladder wall without perforation or damage to adjacent organs. In the case of very large tumors, electrocautery resection is performed first; the base of the tumor is then treated with Nd:YAG laser energy to ensure coagulation necrosis. Thousands of patients have been treated by this method with satisfactory results. The same methods are applicable for small tumors of the ureter. One of the disadvantages of the method is the lack of material for histological examination after the procedure.

HPD–PDT has also been used clinically for therapy of superficial bladder tumors (Marcus, 1992) that do not extend deep into the muscular layers of the bladder. The red laser light penetrates several millimeters into these tumors and efficiently triggers the HPD. The bladder may be distended using transparent fluid or a transparent balloon that is inflated with a fluid. Illumination of the whole bladder is provided by an optical fiber with a spherical diffuser tip (see Section 8.3.2). The preliminary results are encouraging.

9.11.3 Laser Prostatectomy

Endoscopic laser therapy has been tried to for the relief of urethral obstructions due to enlargement of the prostate gland. One of the systems used is transurethral ultrasound-guided laser-induced prostatectomy (TULIP) (Roth and Aretz, 1991). It is based on a rigid stainless steel probe which consists of three parts: ultrasound (US) transducer, Nd:YAG laser catheter, and inflatable balloon. Using ultrasonic guidance, the balloon is placed within the lumen of the prostate and inflated with water. The balloon compresses the prostatic tissue and facilitates the transmission of both US waves and the Nd:YAG laser beam into the tissue. During the procedure the laser beam is used to heat the prostate tissue and to cause coagulation necrosis. The tissue sloughs slowly and creates a larger lumen, relieving the obstruction of the outflow.

9.11.4 Lithotripsy

9.11.4.1 Introduction

Until the 1970s, open surgery was a common way of removing stones from the ureter and the kidneys. A noninvasive technique, called extracorporeal shock wave lithotripsy (ESWL), was introduced for the fragmentation of stones. This method is based on placing a patient in a bathtub and using a spark gap to generate shock waves in the water, which are focused on the stone. The stone fragments may discharge spontaneously. This technique gradually replaced major surgery in cases in which the stones were situated in the upper part of the ureters. Yet, in a significant number of cases the position of the stones or their size makes it impractical to use extracorporeal shock wave lithotripsy. It would be beneficial to fragment the stones endoscopically by a device introduced through the urethra or percutaneously.

9.11.4.2 Endoscopic Lithotripsy

Electrohydraulic Lithotripsy

Shock waves may be generated in water by a miniature spark plug. The electrohydraulic generator contains a special probe which is inserted through a thin urethroscope. It generates shock waves which are used to shatter ureter or bladder stones endoscopically. A major problem of this method is that the instrument shatters large stones into several pieces that have to be crushed mechanically before they are spontaneously discharged. Another problem is the size of the probe. Also, if the electric discharge takes place too close to the ureter wall, it may cause damage.

Ultrasonic Lithotripsy

An ultrasound generator, attached to the tip of a probe, generates waves at a frequency of 20–40 kHz. This probe may be inserted through a urethroscope and placed near a stone. When the tip is placed near a stone immersed in water, it acts like a drill and slowly reduces the stone to small particles. The ultrasonic waves, however, do not affect soft tissue because the impedances of soft tissue and water are almost the same. Some of the problems associated with this method are the relatively large size of the probe and the need to have a fairly high irrigation flow in order to prevent overheating. Although this method is not so effective for bladder stones, it was found to be useful for kidney and ureter stones.

9.11.4.3 Laser Lithotripsy

The two methods described above have several shortcomings; they may cause damage to healthy tissue and both rely on probes that are rather thick and require large endoscopes (10 Fr to 13 Fr). It was therefore suggested in the 1970s that very thin laser endoscopes or laser catheters be used for fragmentation of urethral and bladder stones. Thin endoscopes reduce the risk to the ureter and facilitate improved irrigation flow, while the laser lithotripsy itself causes less damage to tissue. Extensive *in vivo* studies (Dretler, 1988; Steiner, 1988) indicated that the short-pulse Nd : YAG laser and flashlamp-pumped dye laser (see Section 8.6.5) are suitable for this application. Alexandrite lasers have also been used successfully for laser lithotripsy.

Short laser pulses of extremely high peak power are transmitted through

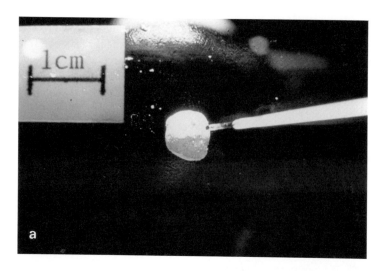

FIGURE 9.18   Laser lithotripsy of stone immersed in water: (a) initial stage; (b)–(d) consecutive stages in the stone fragmentation. (Courtesy of Storz.) (*Figure continues.*)

power fibers and impinge on stones that are immersed in liquids. The pulsed energy generates plasma (accompanied by intense light emission). The plasma, constrained by the surrounding water, generates shock waves that fragment the stones (see Section 3.7.3.2). It was found that the dye lasers generate plasma at lower peak power levels. Nevertheless, many groups prefer the Nd:YAG laser system for lithotripsy.

Figure 9.18 shows the shattering of a stone immersed in water using a pulsed dye laser beam delivered through a fused silica fiber.

FIGURE 9.18   (*Continued*)

It should be mentioned that there are several types of stones with different compositions. Those made of calcium oxalate dihydrate or uric acid are fragile and easy to fragment, whereas those made of calcium oxalate monohydrate,or cystine calculi are not.

Laser Lithotripsy Using Laser Catheters

A thin (e.g., 6 Fr) laser catheter is inserted through the ureter and guided under x-ray control to the immediate vicinity of the stone. The distal tip of the fiber must be placed very close to the stones in order to fragment them in the ureter, bladder, or kidney. If it is placed too far from the tip, the power density will be insufficient to generate shock waves. Some of the methods used to keep the stone close to the tip are shown in Fig. 9.19. In one method, a special basket made of thin, strong metal wires, together with the power fiber, is inserted through the catheter. The basket is used to pull the stone against the fiber tip. Another method involves a balloon catheter; when inflated, the balloon keeps the catheter in position and the fiber is pushed against the stone.

There are several methods for ascertaining that the distal tip of the fiber touches the stone:

(i) Acoustic signals: When shock waves are generated, there are accompanying acoustic signals. By using a stethoscope placed on the abdominal wall of

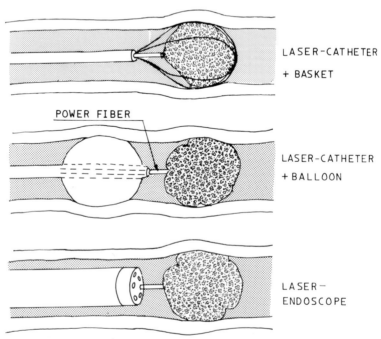

FIGURE 9.19   Laser and fiberoptic systems for laser lithotripsy.

the patient, it is possible to listen to these waves. Alternatively, the acoustic signals generated near the distal tip of the fiber are transmitted back through the power fiber and are detectable at the proximal end of the fiber. These acoustic signals are used for accurate positioning of the fiber.

(ii) Light: The plasma flash consists of a superposition of two emission spectra: a broad spectrum due to the electron transitions in the plasma and a line spectrum due to the emission lines of excited atoms (i.e., calcium). The light emission, transmitted back through the fiber, may easily be measured at the proximal tip of the fiber. The presence of a strong light signal is evidence that the fiber tip is sufficiently close to the stone. It should also be mentioned that the emission spectrum is, in principle, directly correlated to the composition of the stone (calcium oxalate, uric acid, etc.). In practice, while performing stone analysis, it is not easy to find a one-to-one correlation between the emission spectra and the stone composition.

Bhatta *et al.* (1989) studied the possibility of using the acoustic and light signals for "blind" guidance of the laser catheter. They concluded that strong acoustic and strong light signals are produced by calculi. A strong acoustic signal and no light signal suggest that the laser beam is hitting a blood clot. No acoustic and no light signals indicate that the laser is impinging on normal ureter wall or inside the lumen.

During laser lithotripsy, stone particles which are ejected must be washed away; otherwise they will absorb the laser energy and reduce the efficiency of the procedure. Good drainage of the irrigation fluid (i.e., saline solution) is mandatory.

Laser Lithotripsy Using Laser Endoscopes

The first clinical trials involved CW laser beams which were sent through optical fibers. They vaporized the urinary calculi, rather than fragmenting them. This method is likely to cause severe thermal damage and is potentially unsafe. The first experiments with short-pulse Q-switched Nd : YAG lasers were done in the mid-1980s in Germany (Schmidt-Kloiber *et al.*, 1985). They observed the plasma formation when nanosecond pulses of high peak power were focused on stones under water. Later (Hofmann *et al.*, 1988) these researchers reported on the use of this technique for fragmenting urinary calculi.

The first experiments using flashlamp-pumped dye lasers were performed by Watson, Dretler and others (Dretler *et al.*, 1987) Clinical trials were conducted using a 9–11 Fr rigid endoscope (urethroscope) and quartz fibers of diameter 0.2–0.3 mm. The success rate in these trials was high with practically no complications. Later, flexible ureteroscopes were also successfully used. This paved the way for wider clinical use of laser lithotripsy.

Methods identical to those described above may also apply to the fragmentation of kidney (i.e., renal) stones. In this case, the kidney pelvis may be punctured and a rigid endoscope (nephroscope) may be inserted and advanced to the vicinity of the stones. A power fiber is then inserted into the endoscope and used to transmit Nd:YAG or dye laser energy.

9.12 FLOWCHART DIAGRAMS FOR CLINICAL APPLICATIONS OF LASER–FIBER SYSTEMS—ADVANCES

Flowchart diagrams were mentioned in Section 8.7. Here we present a few examples of such diagrams defining clinical procedures. At present there are only few cases in which all the options of laser–fiber system have been applied in a clinical setting. We therefore demonstrate the usefulness of the flowchart diagrams with two *hypothetical* examples. Each procedure which was reported in the literature or which is planned for the future may be presented by a similar diagram.

9.12.1 Flow Diagram for Laser Angioplasty

The flow diagram for a hypothetical laser angioplasty procedure is shown schematically in Fig. 9.20 and the various steps and their implications are described here.

The first step is Start. The next step is imaging, which is normally performed by Angiography. The next step involves decision making; if no blockage is found one goes to Stop; if blockage has been found, one goes on. The next step is again decision making; can one use PTCA for recanalization? If the answer is positive, one goes on to perform PTCA, and if it is successful, the process ends. If the answer is negative, one goes on. The next step is fiberoptic endoscopy accompanied with fiberoptic diagnosis. The following step is decision making regarding whether laser angioplasty could be used for recanalization. If the answer is negative, the process ends. If the answer is positive, one goes on. The next step is choosing the integrated laser–fiber system and the recanalization parameters. The next step is the actual laser angioplasty. Endoscopy and diagnosis are used again to asses the blood flow. The final step involves decision making regarding whether the blood flow is sufficient. If the answer is negative, one has to go back to an earlier step of laser angioplasty. If the answer is positive, the whole process ends.

9.12.2 Flowchart Diagram for Photodynamic Therapy (HPD–PDT)

The diagnosis and therapy steps are shown schematically in Fig. 9.21. Using white light illumination, one starts with regular endoscopy. The next step is decision making. If no tumor is revealed, the process stops. If a tumor is revealed, HPD is injected. The next step usually occurs after 2 days. This step may be fluorescence endoscopy—imaging of the tumor using UV light illumination and imaging through a red filter (alternatively, only the intensity of the red emission is

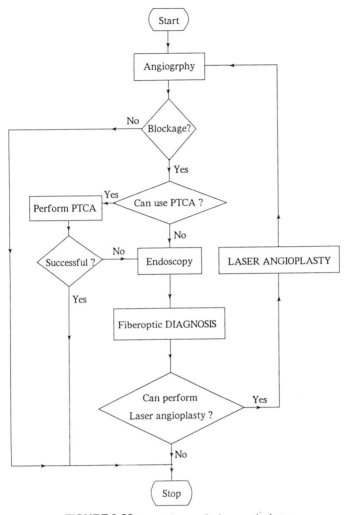

FIGURE 9.20 Flow diagram for laser angioplasty.

measured). The next step is again decision making. If no red emission has been detected, the tumor is benign and the process stops. If red emission has been detected, the tumor is carcinogenic. The area is then illuminated by a red laser. The next step is decision making. Fluorescence endoscopy (or simple detection of the red emission) is used again. If there is no more red emission, the tumor has been eradicated and the process stops. If not, one has to continue with the treatment as shown.

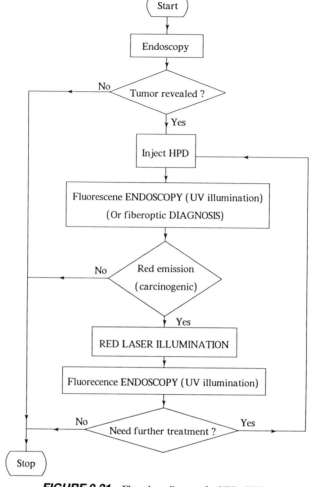

FIGURE 9.21   Flowchart diagram for HPD–PDT.

References

Abela, G. S. (1990). *Lasers in Cardiovascular Medicine and Surgery: Fundamentals and Techniques,* Boston: Kluwer Academic.

Abela, G. S., Normann, S., Cohen, D., Feldman, R. L., Geiser, E. A., and Conti, R. (1982). Effects of carbon dioxide, NdYAG, and argon laser radiation on coronary atheromatous plaques. *Am. J. Cardiol.* **50,** 1199–1205.

Abela, G. S., Seeger, J. M., Barbieri, E., Franzini, D., Fenech, A., Pepine, C. J., and Conti, C. R. (1986). Laser angioplasty with angioscopic guidance in humans. *J. Am. Coll. Cardiol.* **8,** 184–192.

Andersson-Engels, S., Johansson, J., Svanberg, S., and Svanberg, K. (1990). Fluorescence diagnosis and photochemical treatment of deseased tissue using lasers: part II. *Anal. Chem.* **62,** 19a–27a.

Andre, P. J. (1990). Lasers in dermatology. *Ann. Deramatol. Venereol.* **117,** 377–395.

Auer, L. M., Holzer, P., Ascher, P. W., and Heppner, F. (1988). Endoscopic neurosurgery. *Acta Neurochir.* **90**, 1–14.

Baggish, M. S. (1985). *Basic and Advanced Laser Surgery in Gynecology.* Norwalk, CT: Appleton-Century-Crofts.

Baggish, M. S., Sze, E., Badawy, S., and Choe, J. (1988). Carbon dioxide laser laparoscopy by means of a 3.0 mm diameter rigid wave guide. *Fertil. Steril.* **50**, 419–424.

Balchum, O. J., and Doiron, D. R. (1982). Photoradiation therapy of obstructing endobronchial lung cancer. *Proc. SPIE* **357**, 53–59.

Bellina, J. H. (1974). Gynecology and the Laser. *Contemp. Obstet. Gynecol.* **4**, 24–34.

Bellina, J. H., and Bandieramonte, M. D. (1986). *Principles and Practice of Gynecologic Laser Surgery.* New York: Plenum Publishing.

Bellina, J. H., Fick, A. C., and Jackson, J. D. (1985). Lasers in gynecology: an historical/developmental overview. *Laser Surg. Med.* **5**, 1–22.

Bhatta, K. M., Rosen, D., Watson, G. M., and Dretler, S. P. (1989). Acoustic and plasma guided lasertripsy of urinary calculi. *J. Urol.* **142**, 433–437.

Bom, N., and Roelandt, J. (1989). *Intravascular Ultrasound.* Dordrecht: Kluwer Academic Publishers.

Borst, C. (1987). Percutaneous recanalization of arteries: status and prospects of laser angioplasty with modified fibre tips. *Lasers Med. Sci.* **2**, 137–151.

Bown, S. G. (1983). Phototherapy of tumors. *World J. Surg.* **7**, 700–709.

Bown, S. G., Barr, H., Matthewson, K., Hawes, R., Swain, C. P., Clark, C. G., and Boulos, P. B. (1986). Endoscopic treatment of inoperable colerectal cancers with the NdYAG laser. *Br. J. Surg.* **73**, 949–952.

Bridges, T. J., Strand, A. R., Wood, R., Patel, C. K. N., and Karlin, D. B. (1984). Interaction of CO_2 laser radiation with ocular tissue. *IEEE J. Quant. Elect.* **QE-20**, 1449–1458.

Buchelt, M., Papaioannou, T., Fishbein, M., Peters, W., Beeder, C., and Grundfest, W. (1992). Excimer laser ablation of fibrocartilage: an in vitro and in vivo study. *Laser Surg. Med.* **11**, 271–279.

Carruth, J. A. S. (1983). The role of lasers in otolaryngology. *World J. Surg.* **7**, 719–724.

Carruth, J. A. S., and Simpson, G. T. (1988). *Lasers in Otolaryngology.* London: Chapman & Hall.

Cerullo, L. J. (1984). Laser neurosurgery: past, present and future. *IEEE J. Quantum Electron.* **QE-20**, 1397–1400.

Choy, D. S. J., Case, R. B., Fielding, W., Hughes, J., and Liebler, W. (1987). Percutaneous laser nucleolysis of lumbar disks. *N. Engl J. Med.* **317**, 771–772.

Choy, D. S. J., Stertzer, S. H., Myler, R. K., Marco, J., and Fournial, G. (1984). Human coronary laser recanalization. *Clin. Cardiol.* **7**, 377–381.

Cook, A. S., and Rock, J. A. (1991). The role of laparoscopy in the treatment of endometriosis. *Fertil. Steril.* **55**, 663–680.

Cook, S. L., Eigler, N. L., Shefer, A., Goldenberg, T., Forrester, J. S., and Litvack, F. (1991). Percutaneous excimer laser coronary angioplasty of lesions not ideal for balloon angioplasty. *Circulation* **84**, 632–643.

Cragg, A. H., Gardiner, G. A., and Smith, T. P. (1989). Vascular applications of laser. *Radiology* **172**, 925–935.

Daikuzono, N., Suzuki, S., Tajiri, H., Tsunekawa, H., Ohyama, M., and Joffe, S. N. (1988). Laser-thermia: a new computer-controlled contact NdYAG system for interstitial local hyperthermia. *Laser Surg. Med.* **8**, 254–258.

Davis, R. K. (1990). *Lasers in Otolaryngology—Head and Neck Surgery.* Philadelphia: W. B. Saunders.

Doiron, D. R., and Gomer, C. J. (1984). *Porphyrin Localization and Treatment of Tumors.* New York: A. R. Liss.

Douek, P. C., Leon, M. B., Geschwind, H., Cook, P. S., Selzer, P., Miller, D. L., and Bonner, R. F. (1991). Occlusive peripheral vascular disease: a multicenter trial of fluorescence-guided, pulse dye laser–assisted balloon angioplasty. *Radiology* **180**, 127–133.

Dougherty, T. J. (1989). *Photodynamic Therapy: Mechanisms*, SPIE Proceedings, Vol. 1065. Bellingham, WA: SPIE.

Dougherty, T. J. (1990). *Photodynamic Therapy: Mechanisms II*, SPIE Proceedings, Vol. 1203. Bellingham, WA: SPIE.

Dougherty, T. J. (1992). *Optical Methods for Tumor Treatment and Detection*, SPIE Proceedings, Vol. 1645. Bellingham, WA: SPIE.

Dougherty, T. J., Potter, W. R., and Belliner, D. (1990). Photodynamic therapy for the treatment of cancer: current status and advances. In: Kessel, D. (Ed.), *Photodynamic Therapy of Neoplastic Disease*. Boca Raton, FL: CRC Press.

Dretler, S. P. (1988). Laser lithotripsy: a review of 20 years of research and clinical applications. *Laser Surg. Med.* **8**, 341–356.

Dretler, S. P., Watson, G., Parrish, J. A., and Murray, S. (1987). Pulsed dye laser fragmentation of ureteral calculi: initial clinical experience. *J. Urol.* **137**, 386–389.

Dubois, F., Icard, P., Berthelot, G., and Levard, H. (1990). Coelioscopic cholecystectomy. *Ann. Surg.* **211**, 60–62.

Dwyer, R. (1986). The history of gastrointestinal endoscopic laser hemostasis and management. *Endoscopy* **18**, 10–13.

Ellis, S. G., DeCesare, N. B., Pinkerton, C. A., Whitlow, P., King III, S. B., Ghazzal, Z. M. B., Kereiakes, D. J., Popma, J. .J., Menke, K. K., Topol, E. J., and Holmes, D. R. (1991). Relation of stenosis morphology and clinical presentation to the procedural results of directional coronary atherectomy. *Circulation* **84**, 644–653.

Fleischer, D. (1987). Lasers and gastrointestinal disease. In: Sivak, M. V. (Ed.), *Gastroenterologic Endoscopy*, pp. 158–180. Philadelphia: W. B. Saunders.

Forrester, J. S., Eigler, N., and Litvack, F. (1991). Interventional cardiology: the decade ahead. *Circulation* **84**, 942–944.

Fourrier, J. L., Bertrand, M. E., Auth, D. C., Lablanche, J. M., Gommeaux, A., and Brunetaud, J. M. (1989). Percutaneous coronary rotational angioplasty in humans: preliminary report. *J. Am. Coll. Cardiol.* **14**, 1278–1282.

Fruhmorgen, P., Bodem, F., Reidenbach, H. D., Kaduk, B., and Demling, L. (1976). Endoscopic laser coagulation of bleeding gastrointestinal lesions with report of the first therapeutic application in man. *Gastroentintest. Endosc.* **23**, 73–75.

Garrand, T. J., Stetz, M. L., O'Brien, K. M., Gindi, G. R., Sumpio, B. E., and Deckelbaum, L. I. (1991). Design and evaluation of a fiberoptic fluorescence guided laser recanalization system. *Laser Surg. Med.* **11**, 106–116.

Geschwind, H. J., Boussignac, G., Tesseire, B., Vieilledent, C., Gaston, A., Becquemin, J. P., and Mayiolina, P. (1984). Percutaneous transluminal laser angioplasty in man. *Lancet* April, 844.

Ginsburg, R., Wexler, L., Mitchell, R. S., and Profitt, D. (1985). Percutaneous transluminal laser angioplasty for treatment of peripheral vascular disease. *Radiology* **156**, 619–624.

Gomer, C. J., Rucker, N., Ferrario, A., and Wong, S. (1989). Properties and applications of photodynamic therapy. *Radiat. Res.* **120**, 1–18.

Goodale, R. L (1970). Rapid endoscopic control of bleeding gastric erosions by laser radiation. *Arch. Surg.* **101**, 211–214.

Hallock, G. G., and Rice, D. C. (1989). In utero fetal surgery using a milliwatt carbon dioxide laser. *Laser Surg. Med.* **9**, 482–484.

Henderson, B. W and Dougherty, T. J. (1992). *Photodynamic Therapy*. New York: Marcel Dekker.

Hofmann, R., Hartung, R., Geissdorfer, K., Ascherl, R., Erhardt, W., Schmidt-Kloiber, H., and Reichel, E. (1988). Laser induced shock wave lithotripsy—biological effects of nanosecond pulses. *J. Urol.* **139**, 1077–1079.

Hofstetter, A. (1986). Lasers in urology. *Laser Surg. Med.* **6**, 412–414.

Hofstetter, A., and Frank, F. (1980). *The Nd:YAG Laser in Urology*. Basel: Hoffman–LaRoche.

Hunter, J. G. (1989). Endoscopic laser applications in the gastrointestinal tract. *Surg. Clin. North Am.* **69**, 1147–1166.

Hunter, J. G. (1991). Laser or Electrocautery for Laparoscopic Cholecystectomy? *The American Journal of Surgery* **161**, 345–349.

Isner, J. M., and Clarke, R. H. (1984). The current status of lasers in the treatment of cardiovascular disease. *IEEE J. Quantum Electron.* **QE-20**, 1406–1419.

Isner, J. M., Steg, P. G., and Clarke, R. H. (1987). Current status of cardiovascular laser therapy, 1987. *IEEE J. Quantum Electron.* **QE-23**, 1756–1771.

Jain, K. K. (1983). *Handbook of Laser Neurosurgery.* Springfield, IL: Charles C Thomas.

Jain, K. K. (1984). Current status of laser applications in neurosurgery. *IEEE J. Quantum Electron.* **QE-20**, 1401–1406.

Joffe, S. N. (1989). *Lasers in General Surgery.* London: Williams & Wilkins.

Kessel, D. (1990). *Photodynamic Therapy of Neoplastic Diseases.* Boca Raton, FL: CRC Press.

Keye, W. R. (1990). *Laser Surgery in Genecology and Obstetrics*, 2nd ed. Chicago: Year Book Medical Publishers.

Keye, W. R., and Dixon, J. (1983). Photocoagulation of endometriosis by the argon laser through the laparoscope. *Obstet. Gynecol.* **62**, 383–386.

Kiefhaber, P., Nath, G., and Moritz, K. (1977). Endoscopial control of massive gastrointestinal hemorrhage by irradiation with a high power Nd:YAG laser. *Prog. Surg.* **15**, 140–146.

L'Esperance, F. A. (1983). *Ophthalmic Lasers.* St. Louis: C. V. Mosby.

Lee, G., Ikeda, R., Stobbe, D., Ogata, C., Theis, J., Hussein, H., and Mason, D. T. (1983). Laser irradiation of human atherosclerotic obstructive disease: simultaneous visualization and vaporization achieved by a dual fiberoptic catheter. *Am. Heart J.* **105**, 163–164.

Linnemeier, T. J., and Cumberland, D. C. (1989). Percutaneous laser coronary angioplasty without balloon angioplasty. *Lancet* **1**, 155–156.

Lipson, R. L., Baldes, E. J., and Olsen, A. M. (1961). Hematoporphyrin derivative: a new aid for endoscopic detection of malignant disease. *J. Thorac. Cardiovasc. Surg.* **42**, 623–629.

Lipson, R. L., Pratt, J. H., Baldes, E. J., and Dockerty, M. B. (1964). Hematoporphyrin derivative for detection of cervical cancer. *Obstet. Gynecol.* **24**, 78–83.

Litvack, F. (1992). *Coronary Laser Angioplasty.* Boston: Blackwell Scientific Publications.

Lux, G., Ell, C., Hochberger, J., Muller, D., and Demling, L. (1986). The first succesful endoscopic retrograde laser lithotripsy of common bile duct stones in man using a pulsed neodymium-YAG laser. *Endoscopy* **18**, 144–145.

Macruz, R., Martins, J. R. M., Tupinambas, H. S., Lopes, E. A., Penna, A. F., Carvalho, V. B., Armelin, E., and Decourt, L. V. (1980). Possibilidades terapeuticas do raio laser em ateromas. *Arq. Bras. Cardio.* **34**, 9.

Marcus, S. L. (1992). Photodynamic therapy of human cancer. *Proc. IEEE* **80**, 869–889.

Margolis, T. I., Farnath, A., Destro, M., and Puliafito, C. A. (1989). Er:YAG laser surgery on experimental vitreous membranes. *Arch. Opthamol.* **107**, 424–428.

Marshall, G. W., Lipsey, M. R., Heuer, M. A., Kot, C., Smarz, R., and Epstein, M. (1981). An endodontic fiber optic endoscope for viewing instrumented root canals. *J. Endodont.* **7**, 85–88.

McLaughlin, D. S. (1987). Endoscopic laser surgery in gynecology. In: Shapshay, S. M. (Ed.), *Endoscopic Laser Surgery Handbook*, pp. 325–354. New York: Marcel Dekker.

McLaughlin, D. S. (1991). *Lasers in Gynecology.* Philadelphia: J. B. Lippincott.

McNicholas, T. A. (1990). *Lasers in Urology.* New York: Springer-Verlag.

Meyer Schwickerath, G. (1949). Koagulation der Netzhaut mit Sonnenlicht. *Ber. Dtsch. Ophthalmol. Ges.* **55**, 256–259.

Meyers, S. M., Bonner, R. F., Rodrigues, M. M., and Ballintine, E. J. (1983). Phototransection of vitreal membranes with the carbon dioxide laser in rabbits. *Ophthalmology* **90**, 563–568.

Michaels, J. A. (1990). Percutaneous arterial recanalization. *Br. J. Surg.* **77**, 373–379.

Midda, M., and Renton-Harper, P. (1991). Lasers in dentistry. *Br. Dent. J.* **68**, 343–346.

Miller, J. B., Smith, M. R., and Boyer, D. S. (1981). Miniaturized intraocular carbon dioxide laser photosurgical system for multi-incision vitrectomy. *Ophthalmology* **88**, 440–442.

Mizuno, K., Arai, T., Satomura, K., Shibuya, T., Arakawa, K., Okamoto, Y., Miyamoto, A., Kurita, A., Kikuchi, M., Nakamura, H., Utsumi, K., and Takeuchi, K. (1989). New percutaneous transluminal coronary angioscope. *J. Am. Coll. Cardiol.* **13**, 363–368.

Morstyn, G., and Kayo, A. H. (1990). *Phototherapy of Cancer*. New York: Harwood Academic.

Nath, G., Gorisch, W., Kreitmair, A., and Kiefhaber, P. (1973). Transmission of a powerful argon laser beam through a fiberoptic flexible gastroscope for operative gastroscopy. *Endoscopy* **5**, 213–215.

Otsuki, T., Yoshida, R., Miyazawa, T., and Yoshimoto, T. (1992). Endoscopic laser surgery for intraparenchymal brain lesions with computed stereotactic guiding system. *Proc. SPIE* **1649**, 241–243.

Papazoglou, T. G., Papaioannou, T., Arakawa, K., Fishbein, M., Marmarelis, V., and Grundfest, W. (1990). Control of excimer laser aided tissue ablation via laser-induced fluorescence monitoring. *Appl. Opt.* **29**, 4950–4955.

Reddick, E. J., and Olson, D. O. (1989). Laparoscopic laser cholecystectomy: a comparison with mini lap cholecystectomy. *Surg. Endoscop.* **3**, 131–133.

Reis, G., Pomerantz, R., Jenkins, R., Kuntz, R., Baim, D., Diver, D., Schnitt, S., and Safian, R. (1991). Laser balloon angioplasty: clinical, angiographic and histologic results. *J. Am. Coll. Cardiol.* **18**, 193–202.

Robertson, J. H., and Clark, W. C. (1988). *Lasers in Neurosurgery*. Boston: Kluwer.

Ross, D., and Zeidler, G. (1966). Pumping new life into ruby lasers. *Electronics* **39**, 115–118.

Roth, R. A., and Aretz, H. T. (1991). Transurethral ultrasound-guided laser-induced prostatectomy (Tulip procedure): a canine prostrate feasibility study. *J. Urol.* **146**, 1128–1135.

Sanborn, T. A. (1988). Laser angioplasty—what has been learned from experimental studies and clinical trials? *Circulation* **78**, 769–774.

Sanborn, T. A. (1989). *Laser Angioplasty*. New York: A. R. Liss.

Schmidt-Kloiber, H., Reichel, E., and Schoffmann, H. (1985). Laser induced shock wave lithotripsy (LISL). *Biomed. Technik* **30**, 173–181.

Shapshay, S. M. (1987). *Endoscopic Laser Surgery Handbook*. New York: Marcel Dekker.

Sherk, H. H. (1990). *Lasers in Orthopaedics*. Philadelphia: J. B. Lippincott.

Sivak, M. V. (1987). *Gastroenterologic Endoscopy*. London: W. B. Saunders.

Smith, J. A. (1986). Treatment of invasive bladder cancer with a Nd:YAG laser. *J. Urol.* **135**, 55.

Smith, J. A., Stein, B. S., and Benson, R. C. (1990). *Lasers in Urologic Surgery*. Chicago: Year Book Medical Publishers.

Spears, J. R. (1986). Percutaneous laser treatment of atherosclerosis: an overview of emerging techniques. *Cardiovasc. Intervent. Radiol.* **9**, 303–312.

Stack, R. S., Philips, H. E., Quigley, P. J., Tcheng, J. E., Bauman, R. P., Ohman, E. M., O'Neil, W. W., and Walker, C. (1991). Multicenter registry of percutaneous coronary rotational ablation using the rotoblator. *J. Am. Coll. Cardiol.* **17**, 31A.

Staehler, G., Hofstetter, A., Gorisch, W., Keiditsch, E., and Mussiggang, M. (1976). Endoscopy in experimental urology using an argon laser beam. *Endoscopy* **8**, 1–4.

Steger, A. C. (1991). Interstitial laser hyperthermia for the treatment of hepatic and pancreatic tumours. *Photochem. Photobiol.* **53**, 837–844.

Steiner, R. (1988). *Laser Lithotripsy*. New York: Springer-Verlag.

Stern, R. H., and Sognnaes, R. F. (1964). Laser beam effect on dental hard tissues. *J. Dent. Res.* **43**, 73.

Strong, M. S., Jako, G. J., Polanyi, T. G., and Wallach, R. A. (1973). Laser surgery in the aerodigestive tract. *Am. J. Surg.* **126**, 529–533.

Thompson, K. P., Ren, Q. S., and Parel, J. M. (1992). Therapeutic and diagnostic application of lasers in ophthalmology. *Proc. IEEE* **80**, 838–860

Tobis, J. M., Mallery, J. A., Gessert, J., Griffith, J., Mahon, D., McLeay, L., McRae, M., and Henry, W. L. (1989). Intravascular ultrasound cross sectional arterial imaging before and after balloon angioplasty in vitro. *Circulation* **80**, 873–882.

Tschepe, J., Gundlach, P., Leege, N., Hopf, J., Muller, G., and Scherer, H. (1992). The endoscopically controlled laser lithotripsy of salivary gland calculi and the problem of fiber wear. *Proc. SPIE* **1649**, 254–263.

Uchida, Y., Fujimori, Y., Hirose, J., and Oshima, T. (1992). Percutaneous coronary angioscopy. *Jpn. Heart J.* **33**, 271–294.

Waller, B. F. (1989). "Crackers, breakers, stretchers, drillers, scrapers, shavers, burners, welders and melters"—the future treatment of atherosclerotic coronary artery disease? *J. Am. Coll. Cardiol.* **13**, 969–987.

Whipple, T. L. (1987). Orthopedic surgery. In: Shapshay, S. M. (Ed.), *Endoscopic Laser Surgery Handbook*, pp. 397–422. New York: Marcel Dekker.

White, R. A., and Grundfest, W. S. (1989). *Lasers in Cardiovascular Disease*, 2nd ed. Chicago: Year Book Medical Publishers.

Willenberg, G. C. (1989). Dental laser applications: emerging to maturity. *Laser Surg. Med.* **9**, 309–313.

Wollenek, G., Laufer, G., and Grebenwoger, F. (1988). Percutaneous transluminal excimer laser angioplasty in total peripheral artery occlusion in man. *Laser Surg. Med.* **8**, 464–468.

Appendix

A.1 PHYSICS CONSTANTS

Speed of light $c = 3 \times 10^8$ m/sec

Electron charge $e = 1.6 \times 10^{-10}$ coulombs

Planck's constant $h = 6.6 \times 10^{-34}$ joule sec

1 eV $= 1.6 \times 10^{-19}$ joule

A.2 PROPERTIES OF LIGHT

A.2.1 Division of the Optical Spectrum

The optical spectrum may be divided as follows (see Fig. A.1):

Extreme UV: 10–100 nm

Far UV: 100–300 nm

Near UV: 300–390 nm

Visible: 390–780 nm

Near IR: 780 nm–1.5 μm

Middle IR: 1.5–10 μm

Far IR: 10–100 μm

Comment: This division serves just as a guideline. There are authors who use a slightly different division.

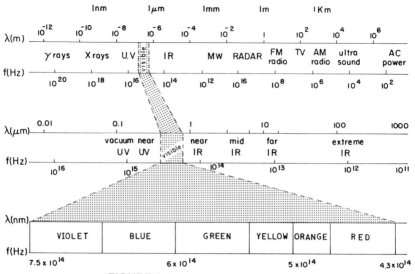

FIGURE A.1 The electromagnetic spectrum.

The wavelengths of the various visible colors are as follows:

Violet: 390–455 nm

Blue: 455–492 nm

Green: 492–577 nm

Yellow: 577–597 nm

Orange: 597–622 nm

Red: 622–780 nm

A.2.2 Light and Radiometry—Terms and Units

From the classical point of view, light consists of electromagnetic radiation; the scientific measurement related to this radiation is called radiometry. Radiometric terms and units are discussed below. Prior to this discussion is a description of the physical quantities and units of energy and power which are commonly used.

Energy Terms

Energy: Measures the amount of work a system is capable of performing. The units which are commonly used are calories or joules. A calorie (cal) is the amount of energy needed to raise the temperature of a gram of water by 1 degree centigrade, and 1 joule (J) is 0.24 calories. Another unit which is commonly used is the electron volt (eV), where $1 \text{ eV} = 1.6 \times 10^{-19} \text{ J}$.

Power: A measure of the rate at which energy is used. The commonly used units are watts; 1 watt = 1 joule/second.

Radiometric Terms

Radiant energy: The energy E carried by a beam of light is described as radiant energy and is measured in joules (J).

Radiant power (flux): The power P is the energy carried by the beam in 1 second and is measured in watts (W), that is, J/sec. It is also defined as radiant flux.

Power density: The power p incident on a unit area. It is measured in W/cm². If power P is incident on an area A, then $p = P/A$. This quantity is also defined as *irradiance* and is particularly important when a beam of light is incident on a surface.

Fluence: The total energy incident on a unit area; it is measured in joules.

Radiant intensity: This the power I emitted by a point source into a unit solid angle; it is measured in W/steradian. This radiant intensity is particularly important in describing light sources such as incandescent lamps.

Comment: In older literature, irradiance was called *intensity*; the term is used loosely in the laser literature. It may be used when there is no danger of confusing it with irradiance.

A few examples will illustrate the various quantities and units:

EXAMPLE I: If a laser beam of energy 75 calories is totally absorbed in 1 g of water at room temperature ($T = 25°C$), its temperature will increase from 25°C to 100°C. Clearly, the same amount of energy will raise the temperature of 75 g of water by 1°C to 26°C.

EXAMPLE II: In a 100-W laser, 100 joules are emitted each second. If the laser operates for 6 sec, the total energy emitted is

$$E = 100 \times 6 = 600 \text{ j} = 144 \text{ cal.}$$

This energy can bring roughly 2 grams of water from room temperature to boiling point.

EXAMPLE III: A beam of power P is incident on an area A for time t.

The irradiance (or power density) is P/A.

The total energy delivered to the area is $E = Pt$.

The fluence is $F = E/A = Pt/A$.

Light and Photons

From the point of view of quantum theory, light consists of a stream of particles called photons. For monochromatic light, the *energy e* of each photon is related to the frequency v by the famous formula $e = hv$, where h is Planck's constant ($h = 6.6 \times 10^{-34}$ J/sec). Alternatively, one could write $e = hc/\lambda$ where c is the light velocity ($c = 3 \times 10^8$ m/sec) and λ is the wavelength. The photon energy is inversely proportional to wavelength. Therefore a photon of UV light is more energetic than a photon of visible light, and the latter is more energetic than a photon of IR light.

The *energy E* in a beam of light is the sum total of the energies of the photons; that is, it is the energy *e* multiplied by the total number of photons N, $E = eN$.

The *power P* in the monochromatic beam is determined by the total energy $E = eN$ that passes through an area in a unit time, $P = eN/t$. If the beam is incident on a surface area A, the power density on the surface is again given by P/A.

EXAMPLE I. *CW Lasers*: For an HeNe laser the wavelength is $\lambda = 633$ nm with emission in the red; the frequency is $\nu = 4.8 \times 10^{14}$ Hz and the energy per photon is $e_r = 3 \times 10^{-19}$ J $= 1.9$ eV. For a 1-mW HeNe laser an energy of 10^{-3} J is emitted per second. If we denote by N_r the number of photons that are emitted per second, then $N_r \times e_r = 1$ mW, that is, $N_r = 3.3 \times 10^{15}$ photons per second.

For an HeCd laser the wavelength is $\lambda = 325$ nm with emission in the UV; the frequency is $\nu = 9.35 \times 10^{14}$ Hz and the energy per photon is $e_{uv} = 5.9 \times 10^{-19}$ J $= 3.7$ eV. For a 1-mW HeCd laser an energy of 10^{-3} J is emitted per second. In this case $N_{uv} = 1.7 \times 10^{15}$ photons per second.

In the red beam, the individual photons have lower energies e_r than the photon energy e_{uv} in a UV beam. On the other hand the total number of red photons emitted per second is larger than the corresponding number of UV photons, so that $N_r e_r = N_{uv} e_{uv}$.

An infrared laser (i.e., CO_2) beam may consist of a very large number of photons, each of which has a low energy (e.g., 0.2 eV). The total energy or power density of such a beam may be high and should not be confused with the individual photon energy in this beam!

EXAMPLE II. *Pulse Lasers*: For an XeCl excimer laser the wavelength is $\lambda = 308$ nm; the frequency is $\nu = 9.7 \times 10^{14}$ Hz and the energy per photon is 6.4×10^{-19} J $= 4$ eV. If the laser emits 10 mJ per pulse, there are 1.5×10^{16} photons in that pulse.

A.3 REFLECTION AND ABSORPTION OF LIGHT IN A SAMPLE

Let us consider a beam of intensity I_0 that is normally incident on a slab of thickness t, shown in Fig. 3.1, and define $K = \alpha\lambda/4\pi$. In the noncoherent case, the reflectance is

$$r = I_r/I_0 = [(n - 1)^2 + K^2] / [(n + 1)^2 + K^2]$$

and the transmittance is

$$\tau = I_t/I_0 = [(1 - r)^2 \exp(-\alpha t)] / [1 - r^2 \exp(-2\alpha t)].$$

When there is no absorption, $\alpha = 0$, one obtains the formulas used in Section 3.3.1

Glossary

Ablation Removal of tissue.

Absorbance (or absorptance) Ratio of the absorbed light intensity I_a to the incident intensity I_i. It is a dimensionless quantity.

Absorption The transformation of light (radiant) energy to some other form of energy—usually heat—as the light traverses matter.

Absorption coefficient In a nonscattering sample, the reciprocal of the distance l over which light of intensity I is attenuated (due to absorption) to $I/e \approx I/3$. The units are typically cm^{-1}.

Acceptance angle Maximum incident angle at which an optical fiber will transmit light by total internal reflection.

Anastomosis Connection of two tubes. In particular, laser heat facilitates anastomosis of blood vessels.

Angiography Use of x-ray imaging to reveal disease in arteries. A radiopaque contrast medium (angiographic dye) injected into arteries absorbs the x-rays. The dye is seen as dark lines on the x-ray fluorescence screen.

Angioplasty The "reshaping" of blood vessels.

Angioscopy Endoscopic imaging inside the blood vessels in the cardiovascular system.

Ar (argon) ion laser A laser with a lasing medium composed of ionized argon gas; the emission is in the visible ($0.5-0.6 \ \mu m$).

Articulated arm An assembly consisting of several mirrors that are mounted on hinges. The assembly is used to direct a beam from the laser head to the target tissue.

Atheroma Fatty degeneration of the inner walls (intima) of the arteries in arteriosclerosis.

Atherectomy Excision of atheroma.

Atherosclerotic plaque A fibrous tissue that also contains fat and sometimes calcium. It accumulates in arteries and produces occlusion.

Attenuation   Decrease in the intensity of light passing through matter. Attenuation is caused by reflection, scattering, and absorption.

Attenuation coefficient   Reciprocal of the distance l over which light of intensity I is attenuated to $I/e \approx I/3$. The units are typically cm^{-1}.

Autofluorescence   Natural tissue fluorescence.

Beam   A slender stream of light.

Balloon catheter   A catheter that includes an inflatable balloon at its end.

Bifurcated bundle of fibers   A bundle of fibers that is divided into two branches on one end.

Biliary   Pertaining to the bile.

Biliary calculus   Gallbladder stone.

Biocompatible   Something that does not cause harm to biological tissue.

"Bypass" operation   See CABG

CABG (coronary artery bypass grafting)   An open heart operation in which a section of a vein or an artery is used to bypass a blocked coronary artery.

Calorie   A unit of energy. One calorie is the amount of work needed to raise the temperature of 1 gram of water at 15°C by 1 degree centigrade. 1 calorie = 4.2 joules.

Cannula   A tube which is inserted into the body. It is often fitted with a pointed rod for ease of insertion.

Cataract   A condition in which the lens of the eye becomes opaque, causing partial blindness.

Catheter   A flexible hollow tube normally employed to inject liquids or drain fluids from body cavities.

CCD (charged-coupled device) camera   A solid-state electronic device that serves as an imaging chip and is used in miniature video cameras.

Chemical F/O sensors   Fiberoptic sensors that detect chemical parameters (e.g., pH).

Cholecystectomy   Surgical removal of the gallbladder (*chole*—gall; *cystectomy*— excision of a bladder).

Chromophore   Any coloring agent in tissue which absorbs light. The main chromophore in skin is melanin.

Cladding   Outer part of an optical fiber. It has a lower refractive index than the core.

CO (carbon monoxide) laser   A laser where the lasing medium is CO gas with IR emission at 5 μm.

CO$_2$ (carbon dioxide) laser   A laser where the lasing medium is CO$_2$ gas with IR emission at 10.6 μm.

Coagulation   Change of state of a fluid (e.g., blood) from liquid to semisolid. In the case of exposure to laser beams, coagulation results from overheating.

Collimated beam   Beam of light in which all the rays are parallel to each other.

CW (continuous wave)   Continuous operation, with no interruptions (e.g., of a laser).

Core   Inner part of an optical fiber through which light propagates. It has a higher refractive index than the cladding.

Cornea A transparent watchglass-like tissue that covers the front of the eye.

Coronary artery One of the major arteries that supply blood to the heart.

Coupler An optical device that interconnects optical components.

Critical angle Minimum incidence angle in a medium where light is totally internally reflected.

CT (computed tomography) A noninvasive x-ray imaging method that is particularly useful for the detection and for obtaining three-dimensional images of tumors inside the body.

Cyst A saclike structure or a pocket in the body; it is often filled with a fluid.

Cytotoxic Having toxic effects on cells (and tissue).

dB (decibel) Engineering unit for the ratio of the input power P_{in} in a given device to the output power P_o. It is convenient to measure the logarithm of the ratio $\log(P_o/P_{in})$, and the dB is a standard unit that is equal to 10 times that log; 1 dB = $10 \log(P_o/P_{in})$.

Depth of field The correct term is depth of focus (see below).

Depth of focus Axial distance over which a beam is clearly focused. In endoscopy, the axial distance over which the image is still clearly focused.

Dichroic A surface that reflects different colors when viewed in different directions.

Direct F/O sensor A fiberoptic sensor without an optode attached to the distal tip of the fiber.

Divergence The "spreading" of a light beam in general, and in particular of a laser beam as it moves away from the laser.

Dye laser A laser where the laser medium is a liquid dye. Dye lasers emit in a broad spectral range (e.g., in the visible) and are tunable.

Diskectomy Removal of a disk that causes pressure and pain in the spine.

Efficiency The overall efficiency of a laser is the ratio of the input electrical energy to the output energy in the laser beam.

EM (electromagnetic) radiation Flow of energy that is related to the vibration of electric and magnetic waves.

EM (electromagnetic) spectrum The entirety of the electromagnetic waves that differ from each other in frequency and wavelength.

Emission spectrum The emission obtained from a luminescent material at different wavelengths, when excited by a narrow range of shorter wavelengths.

Endarterectomy Removal of diseased endothelium from within a blood vessel.

Endoscope Optical instrument used for viewing internal organs.

Energy The product of power (watts) and time (sec). Energy is measured in joules (J).

Er:YAG (erbium:yttrium aluminum garnet) laser A solid-state laser whose lasing medium is the crystal Er:YAG with emission in the mid-IR at 2.94 μm.

Excimer laser A laser whose lasing medium is an excited molecular complex (e.g., KrF or ArF); the emission is in the UV ($\lambda < 400$ nm).

Excitation spectrum The emission spectrum at one wavelength is monitored and the intensity at this wavelength is measured as a function of the exciting wavelength.

Extinction length   Distance over which light is attenuated in an absorbing material by a factor of 100.

F number (*f*#)   Ratio of the focal length *f* of a lens to its diameter *D*; $f\# = f/D$.

FEL (free-electron laser)   A laser that is based on the emission from accelerated electrons. The laser is tunable over a wide spectral range.

Fiberscope   A viewing instrument that incorporates an ordered bundle for imaging and a light guide for illumination.

Field of view   The extent of an object that can be imaged or seen through an optical system.

Fluence   The total energy that is incident on a unit area. It is a product of the power density of the laser beam and the irradiation time.

Fluorescence   Luminescence that essentially occurs simultaneously with the excitation of a sample.

F/O   Fiberoptic.

Focal spot   The spot obtained at the focus of a lens. The size of the spot depends on the lens and on the wavelength, but its diameter is never smaller than the wavelength of light.

Fr (French)   Measure of the diameter of a medical catheter; 1 Fr = 1/3 mm.

Fresnel reflection (or Fresnel loss)   When light travels from one medium to a second medium with a different index of refraction, part of the light is transmitted into the second medium and part is reflected. The reflection is referred to as Fresnel reflection or Fresnel loss.

GaAs laser   A laser based on the semiconductor material GaAs. The emission is in the near IR, at about 1 μm.

Gaussian beam   If the intensity at the center of the beam is I_0, then the formula for a Gaussian beam is $I = I_0 \exp(-2r^2/w^2)$ where r is the radial distance from the axis and w is the beam "waist." The intensity profile of such a beam is said to be bell shaped.

Glaucoma   A disease of the eye characterized by abnormal interocular pressure that may lead to loss of sight.

Glioma   Neoplasm derived from cells in the brain, spinal chord, and pituitary gland.

Guide wire   A flexible wire that is inserted into the body and threaded through the vascular system. The wire may then be used to guide a catheter or an endoscope to a desired location.

HeNe (helium neon) laser   A gas laser whose laser medium is a mixture of the gases He and Ne; the emission is in the red (0.628 μm).

Hertz (Hz)   A unit of frequency that is equal to 1 cycle per second. It is often used to indicate the pulse repetition rate of a laser (e.g., a 10-Hz laser emits 10 pulses per second).

HF (hydrogen fluoride) laser   A gas laser whose laser medium is the gas HF and whose emission is in the IR (with several lines between 2.7 and 2.9 μm).

HPD (hematoporphyrin derivative)   A compound used in cancer diagnostics and therapy.

Hyperplasia   Increase in the size of tissue or organ.

Image guide   An ordered bundle of fibers that is used for image transmission.

Index of refraction   Ratio of the velocity of light in a vacuum to the velocity of light in a given material.

Indirect sensor   A fiberoptic sensor whose optode is attached to the end of the fiber.

Infrared (IR)   The part of the electromagnetic spectrum that is invisible and extends between 0.7 and 1000 μm.

Intensity (i.e., radiant intensity)   Power emitted into a unit solid angle (watts/steradian W/sr).

Intima   The inner wall of a vessel.

Intraluminal   Within the lumen of a cylindrical organ.

In vitro   Inanimate matter (in medicine, pertaining to experiments on dead tissue).

In vivo   Of living matter (in medicine, pertaining to experiments on animals or humans).

Irradiance   Ratio of the power incident on a sample to the illuminated area.

Joule   A unit of energy that is equal to 0.24 calorie.

Laparoscope   An instrument that is introduced into the abdomen; it is normally used for imaging the pelvic organs.

Laser   Acronym for light amplification by the stimulated emission of radiation. A device that generates a beam of light that is collimated, monochromatic, and coherent.

Laser angioplasty   Use of laser beams sent through power fibers for the removal of blockages in arteries.

Laser catheter   A catheter that incorporates an optical fiber for the transmission of a laser beam.

Laser endoscope   An endoscope that incorporates an optical fiber for the transmission of a laser beam.

Laser power   Rate of radiation emission from a laser, normally expressed in watts (W).

LDV (laser Doppler velocimeter)   A laser technique for measuring the velocity of a moving body. It may be used for measuring the velocity distribution of the blood cells (and thus the blood flow).

Lesion   An injury or an alteration of an organ or tissue.

Light guide   Assembly of optical fibers that are bundled but not ordered (noncoherent) and are used for illumination.

Lithotripsy   Shattering of stones in the body. Laser lithotripsy refers to shattering by the application of an intense laser beam.

Lumen   The passage contained within the walls of a tube (in particular, the opening inside a blood vessel or the opening inside a catheter).

Luminal   A property which applies to the lumen.

Luminescence   Light emitted from a sample which is irradiated (excited) by energetic photons.

Lysis   Gradual and successful destruction of cells (or ending of a disease).

Micrometer (i.e., micron or μm)   A unit of length that is equal to a thousandth of a millimeter.

Microsecond (μsec)   One-millionth of a second.

Microwave   Electromagnetic waves in the frequency range $10^9 - 10^{11}$ Hz.

Millijoule (mJ)   One-thousandth of a joule.

Millisecond (msec)   One-thousandth of a second.

Monochromatic   Of one color only (e.g., a laser beam of one color); in practice, a beam containing a very narrow range of wavelengths.

MRI (magnetic resonance imaging)   A noninvasive imaging technique that is based on magnetic resonance methods. It provides a wealth of information about inner structures in the body and in particular about tumors.

Multifiber   A small bundle of fibers.

MW (megawatt)   One million watts (10^6 watts).

mW (milliwatt)   One-thousandth of a watt (10^{-3} watt).

Myocardium   Heart muscle tissue.

nm (nanometer)   A unit of length equal to one-billionth of a meter (10^{-9} m), or one-millionth of a millimeter.

nsec (nanosecond)   One billionth of a second (10^{-9} second).

Nd:YAG (neodymium:yttrium aluminum garnet) laser   A solid-state laser whose lasing medium is the crystal Nd:YAG with emission in the near IR, at 1.06 μm.

Necrosis   Death or decay of tissue.

Normal incidence   Incidence of a light beam on a plane at an angle of 90° to the plane.

Numerical aperture (NA)   Light-gathering power of an optical fiber. It is proportional to the sine of the acceptance angle.

Optical detector   A device that converts optical energy to an electrical signal.

Optical fiber   Thin and transparent thread through which light can be transmitted by total internal reflection.

Optical filter   A device that transmits only part of the spectrum incident on it.

Optode   A transducer that is attached to the distal tip of a fiberoptic sensor. The interaction between the optode and the body is monitored by the fiberoptic sensor.

Ordered (coherent) bundle   Assembly of optical fibers that are ordered in exactly the same way at both ends of the bundle.

Palliate   Alleviate pain or disease.

Photon   The fundamental unit of light energy.

Phosphorescence   Luminescence which is delayed with respect to the excitation of a sample.

Photosensitizer   A substance that increases the absorption of another substance at a particular wavelength band.

Physical F/O sensors   Sensors that measure "physical" quantities such as pressure or temperature.

Plaque   See atherosclerotic plaque.

Plasma (physics)   Ionized gas, at high temperature.

Plasma (medicine)   The fluid part of the blood.

Power   The rate of delivery of energy. It is normally measured in watts, that is joules per second.

Power density   The power (e.g., of an incident laser beam) divided by the area on which it is incident. The units are watts/cm^2.

Power fiber   Optical fiber that can transmit a laser beam of relatively high intensity.

PTCA (percutaneous transluminal coronary angioplasty)   A procedure based on a balloon catheter that is inserted into a blocked coronary artery. The lumen is enlarged by inflating the balloon.

Recanalization   See laser angioplasty.

Reflectance (or reflection coefficient)   The ratio of the intensity I_r reflected from a surface to the incident intensity I_i. It is a dimensionless quantity.

Renal   Pertaining to the kidney.

Repetition rate   Number of pulses (e.g., laser pulses) per second. The repetition rate is measured in hertz.

Resolution   Measure of the ability of an optical imaging system to reveal details of an imagel, i.e., to resolve adjacent elements.

RF (radio frequency)   The part of the EM spectrum between about 10^6 and 10^8 hertz.

RT (room temperature)   A temperature of about $27°$ C or 300 K.

Saline solution   A salt solution that is used for medical treatment. This solution is designed to have the same osmotic pressure as blood.

Spectrum   Range of frequencies or wavelengths.

Stenosis   Narrowing.

Stent   A device used to maintain some body orifice open.

TEA (transversely excited atmospheric) CO_2 laser   A special CO_2 gas laser that operates at atmospheric pressure. It emits very short pulses of very high peak power.

Total internal reflection   Reflection of light at the interface between media of different refractive indices, when the angle of incidence is larger than a critical angle (determined by the media).

Transmittance   Ratio of the intensity transmitted through a sample I_t to the incident intensity I_i. It is a dimensionless quantity.

Trocar   A surgical tool that consists of a sharp-ended rod which is enclosed in a wider tube (cannula). The trocar is inserted through the skin into a body cavity and the rod is withdrawn, leaving the tube in place.

Tunable laser   Most lasers emit at a particular wavelength. In tunable lasers, one can vary the wavelength over some limited spectral range.

Ultrasound   Mechanical vibrations with frequencies in the range $2 \times 10^4 - 10^7$ Hz.

UV (ultraviolet)   The part of the optical spectrum that extends between about 10 and 390 nm.

Vacuolation   Creation of spaces or holes in tissue.

Visible   The part of the optical spectrum, roughly in the range 0.4 to 0.7 μm, that can be sensed by the human eye.

Vitreous humor   A transparent jellylike substance that fills the chamber between the lens and the back of the eye.

Watt   Unit of power. One watt is equal to 1 joule per second.

Wavelength   Distance between two adjacent peaks in a wave (e.g., in an EM wave).

Bibliography

Abela, G. S. *Lasers in Cardiovascular Medicine and Surgery: Fundamentals and Techniques*, Boston: Kluwer Academic, 1990.

Allan, W. B. *Fiber Optics: Theory and Practice*, New York: Plenum Publishing, 1973.

Allard, F. C. *Fiber Optics Handbook*, New York: McGraw-Hill, 1990.

Apfelberg, D. B. *Evaluation and Installation of Surgical Laser Systems*, New York: Springer-Verlag, 1987.

Atsumi, K. *New Frontiers in Laser Medicine and Surgery*, Amsterdam: Excerpta Medica, 1983.

Baggish, M. S. *Basic and Advanced Laser Surgery in Gynecology*, Norwalk, CT: Appleton-Century-Crofts, 1985.

Bellina, J. H. and Bandieramonte, M. D. *Principles and Practice of Gynecologic Laser Surgery*, New York: Plenum Publishing, 1986.

Berci, G. *Endoscopy*, New York: Appleton-Century-Crofts, 1976.

Berns, M. W. *Laser Interaction with Tissue*, SPIE Proceedings, Vol. 908, Bellingham, WA: SPIE, 1988.

Berry, M. J. and Harpole, G. M. *Thermal and Optical Interactions with Biological and Related Composite Materials*, SPIE Proceedings, Vol. 1064, Bellingham, WA: SPIE, 1989.

Bom, N. and Roelandt, J. *Intravascular Ultrasound*, Dordrecht: Kluwer Academic Publishers, 1989.

Carruth, J. A. S. and McKenzie, A. L. *Medical Lasers: Science and Clinical Practice*, Bristol, England: Adam Hilger, 1986.

Carruth, J. A. S. and Simpson, G. T. *Lasers in Otolaryngology*, London: Chapman & Hall, 1988.

Cherin, A. H. *An Introduction to Optical Fibers*, New York: McGraw-Hill, 1983.

Culshaw, B. *Optical Fibre Sensing and Signal Processing*, London: Peter Peregrinus, 1982.

Davis, R. K. *Lasers in Otolaryngology—Head and Neck Surgery*, Philadelphia: W. B. Saunders, 1990.

Doiron, D. R. and Gomer, C. J. *Porphyrin Localization and Treatment of Tumors*, New York: A. R. Liss, 1984.

Dougherty, T. J. *Photodynamic Therapy: Mechanisms*, SPIE Proceedings, Vol. 1065, Bellingham, WA: SPIE, 1989.

Dougherty, T. J. *Photodynamic Therapy: Mechanisms II*, SPIE Proceedings, Vol. 1203, Bellingham, WA: SPIE, 1990.

Dougherty, T. J. *Optical Methods for Tumor Treatment and Detection*, SPIE Proceedings, Vol. 1645, Bellingham, WA: SPIE, 1992.

Driscoll, W. G. and Vaughan, W. *Handbook of Optics*, New York: McGraw-Hill, 1978.

Goldman, L. *Laser Non Surgical Medicine*, Lancaster: Tachnomic, 1991.

Harrington, J. A. *Infrared Fiber Optics*, Bellingham, WA: SPIE, 1990.

Hecht, E. *Optics*, 2nd ed., Reading, MA: Addison-Wesley, 1987.

Hecht, J. *The Laser Guidebook*, 2nd ed., New York: McGraw-Hill, 1991.

Hofstetter, A. and Frank, F. *The Nd:YAG Laser in Urology*, Basel: Hoffman–LaRoche, 1980.

Jacques, S. L. *Laser–Tissue Interaction*, SPIE Proceedings, Vol. 1202, Bellingham, WA: SPIE, 1990.

Jacques, S. L. *Laser–Tissue Interaction II*, SPIE Proceedings, Vol. 1427, Bellingham, WA: SPIE, 1991.

Jacques, S. L. *Laser–Tissue Interaction III*, SPIE Proceedings, Vol. 1646, Bellingham: SPIE, 1992.

Jain, K. K. *Handbook of Laser Neurosurgery*, Springfield, IL: Charles C Thomas, 1983, pp. 6–47.

Joffe, S. N. *Laser Surgery: Chracterization and Therapeutics*, SPIE Proceedings, Vol. 907, Bellingham, WA: SPIE, 1988.

Joffe, S. N. *Lasers in General Surgery*, London: Williams & Wilkins, 1989.

Joffe, S. N. and Atsumi, K. *Laser Surgery: Advanced Characterization, Therapeutics and Systems II*, SPIE Proceedings, Vol. 1200, Bellingham, WA: SPIE, 1990.

Joffe, S. N., Goldblatt, N. R. and Atsumi, K. *Laser Surgery: Advanced Characterization, Therapeutics and Systems*, SPIE Proceedings, Vol. 1066, Bellingham, WA: SPIE, 1989.

Joffe, S. N., Parrish, J. A. and Scott, R. S. *Lasers in Medicine*, SPIE Proceedings, Vol. 712, Bellingham, WA: SPIE, 1986.

Kapany, N. S. *Fiber Optics Principles and Applications*, New York: Academic Press, 1967.

Katzir, A. *Novel Optical Fiber Techniques for Medical Applications*, SPIE Proceedings, Vol. 494, Bellingham, WA: SPIE, 1984.

Katzir, A. *Optical Fibers in Medicine and Biology*, SPIE Proceedings, Vol. 576, Bellingham, WA: SPIE, 1985.

Katzir, A. *Optical Fibers in Medicine II*, SPIE Proceedings, Vol. 713, Bellingham, WA: SPIE, 1986.

Katzir, A. *Optical Fibers in Medicine III*, SPIE Proceedings, Vol. 906, Bellingham, WA: SPIE, 1988.

Katzir, A. *Optical Fibers in Medicine IV*, SPIE Proceedings, Vol. 1067, Bellingham, WA: SPIE, 1989.

Katzir, A. *Optical Fibers in Medicine V*, SPIE Proceedings, Vol. 1201, Bellingham, WA: SPIE, 1990.

Katzir, A. *Optical Fibers in Medicine VI*, SPIE Proceedings, Vol. 1420, Bellingham, WA: SPIE, 1991.

Katzir, A. *Optical Fibers in Medicine VII*, SPIE Proceedings, Vol. 1649, Bellingham, WA: SPIE, 1992.

Keye, W. R. *Laser Surgery in Gynecology and Obstetrics*, 2nd ed., Chicago: Year Book Medical Publishers, 1990.

L'Esperance, F. A. *Ophthalmic Lasers*, St. Louis: C. V. Mosby, 1983.

Levi, L. *Applied Optics*, Vol. 1, New York: Wiley, 1968.

Levi, L. *Applied Optics*, Vol. 2, New York: Wiley, 1980.

Lewandowski, R. *Elektrische Licht in der Heilkunde*, Wien: Urben & Schwarzenberg, 1892.

Litvack, F. *Coronary Laser Angioplasty*, Boston: Blackwell Scientific Publications, 1992.

Martellucci, S. and Chester, A. N. *Laser Photobiology and Photomedicine*, Ettore Majorana Int. Sci. Series, 1989.

McLaughlin, D. S. *Lasers in Gynecology*, Philadelphia: J. B. Lippincott, 1991.

McNicholas, T. A. *Lasers in Urology*, New York: Springer-Verlag, 1990.

Meyers, R. A. *Encyclopedia of Lasers and Optical Technology*, New York: Academic Press, 1991.

Podbielska, H. Holography, Interferometry and Optical Pattern Recognition in Biomedicine II, SPIE Proceedings, Vol. 1647, Bellingham, WA: SPIE, 1992.

Pratesi, R. *Optronic Techniques in Diagnostic and Therapeutic Medicine*, New York: Plenum, 1991.

Pratesi, R. and Sacchi, C. A. *Lasers in Photomedicine and Photobiology*, New York: Springer-Verlag, 1980.

Regan, J. D. and Parrish, J. A. *The Science of Photomedicine*, New York: Plenum, 1982.

Robertson, J. H. and Clark, W. C. *Lasers in Neurosurgery*, Boston: Kluwer, 1988.

Salmon, P. R. *Fibre Optic Endoscopy*, New York: Grune & Stratton, 1974.

Sanborn, T. A. *Laser Angioplasty*, New York: A. R. Liss, 1989.

Seippel, R. G. *Fiber Optics*, Reston VA: Prentice Hall, 1984.

Shapshay, S. M. *Endoscopic Laser Surgery Handbook*, New York: Marcel Dekker, 1987.

Sherk, H. H. *Lasers in Orthopaedics*, Philadelphia: J. B. Lippincott, 1990.

Siegman, A. E. *Lasers*, Mill Valley, CA: University Science Books, 1986.

Sivak, M. V. *Gastroenterologic Endoscopy*, London: Saunders, 1987.

Sliney, D. and Wolbarsht, M. *Safety with Lasers and Other Optical Sources*, New York: Plenum Publishing, 1980.

Smith, J. A., Stein, B. S. and Benson, R. C. *Lasers in Urologic Surgery*, Chicago: Year Book Medical Publishers, 1990.

Steiner, R. *Laser Lithotripsy*, New York: Springer-Verlag, 1988.

West, A. I. *Microsensors and Catheter Based Imaging Technology*, SPIE Proceedings, Vol. 904, Bellingham, WA: SPIE, 1988.

West, A. I. *Catheter Based Sensing and Imaging Technology*, SPIE Proceedings, Vol. 1068, Bellingham, WA: SPIE, 1989.

White, R. A. and Grundfest, W. S. *Lasers in Cardiovascular Disease*, 2nd ed., Chicago: Year Book Medical Publishers, 1989.

White, R. A. and Klein, S. R. *Endoscopic Surgery*, St Louis: Mosby Year Book, 1991.

Wilson, J. and Hawkes, J. F. B. *Lasers Principles and Applications*, New York: Prentice Hall, 1987.

Wilson, J. and Hawkes, J. F. B. *Optoelectronics: An Introduction*, 2nd ed., New York: Prentice Hall, 1989.

Winburn, D. C. *Lasers*, New York: Marcel Dekker, 1987.

Wise, D. L. and Wingard, L. B. *Biosensors with Fiberoptics*, Clifton, NJ: Humana Press, 1991.

Wolf, H. F. *Handbook of Fiber Optics: Theory and Applications*, London: Granada, 1979.

Wolfbeis, O. S. *Fiber Optic Chemical Sensors and Biosensors*, Vol. I, Boca Raton, FL: CRC Press, 1991.

Yariv, A. *Optical Electronics*, 4th ed., Philadelphia: Holt, Reinhart & Winston, 1991.

Index